Project Management Process Improvement

Recent Titles in the Artech House Effective Project Management Series

Robert K. Wysocki, Series Editor

The Project Management Communications Toolkit, Carl Pritchard

Project Management Process Improvement, Robert K. Wysocki

For a listing of recent titles in the *Artech House Technology Management and Professional Development Library*, turn to the back of this book.

Project Management Process Improvement

Robert K. Wysocki

Artech House
Boston • London
www.artechhouse.com

Library of Congress Cataloging-in-Publication Data

A catalog record for this book is available from the Library of Congress.

British Library Cataloguing in Publication Data

A catalog record for this book is available from the British Library.

Cover design by Gary Ragaglia

685 Canton Street
Norwood, MA 02062

International Standard Book Number: 1-58053-717-0

10 9 8 7 6 5 4 3 2 1

Contents

Introduction

Regardless of the extent of your efforts to design and deploy an exemplary project management process, the realities always seem to fall far short of your expectations. Project managers and their teams do not use the processes you worked so hard to get accepted and approved. Some use the processes sporadically. Others modify them to suit their own purposes. Others ignore them all together and opt to use their own processes, which they believe to be superior to the ones offered by their organization. Risk underlies many of these decisions. Project managers would prefer to use the tried and true rather than expose the project to added risk through the use of unfamiliar processes. Only after prompting from you do they give the new processes a try. Their reluctance is obvious and you do not seem to have won any converts to the new way. The question remains: How do you win converts to the processes and support you are offering?

If you are experiencing these behaviors, I do not want you to think that they are due to any failure on your part. Much to the contrary, to have a project management process in place is a considerable accomplishment. You should be proud of having reached this goal. However, I want you to take the position that it is just an intermediate goal. The long-term goal is to have a fully developed and integrated portfolio of project management processes. Your initial offering is just that—initial. You are now dealing with a cultural change. As teams begin to use these processes, you will learn about the process shortcomings. These you will fix, and the cycle of learning and fixing will continue. This improvement cycle is a major focus of this book.

This book will take you through the steps you will need in order to mature your project management processes to whatever level your management expects.

In addition to the process goals, you will want to achieve a certain level of adoption, and we will consider the steps to achieve that goal as well.

The book is organized chronologically and designed to be read from cover to cover. The first three chapters establish the foundational information we will need for the remainder of the book. Chapter 1 sets the stage for process improvement. Its primary contribution is a presentation of the process improvement life cycle. This is the roadmap for the remainder of the book. Chapter 2 establishes the concepts that drive all further discussion. A project management maturity model is introduced as an adaptation of the Software Engineering Institute's Capability Maturity Model® for software development. The nine knowledge areas defined by the Project Management Institute, which are introduced in Chapter 2, are the infrastructure upon which we are going to build a strategy for process improvement. This chapter also introduces the process maturity matrix as the entity on which all improvement initiatives are based. Chapter 3 describes the survey instrument that is used for data collection. It serves two purposes. The first is to set the baseline project management maturity. This is an assessment of the maturity level of the project management processes themselves. This is the baseline against which all projects will be assessed as to the maturity level of the practice of the processes. There will be gaps between the maturity of the processes and the maturity of the practice of the processes. Chapter 3 will address those gaps and the information they convey regarding the organization's project management effectiveness.

Chapters 4 and 5 present the tools that we will use to analyze and present the maturity data and identify improvement opportunities. Chapter 4 presents project level performance tools that are familiar to most project managers: cost schedule control (a.k.a. earned value), milestone trend charts, and project reviews. Chapter 4 closes with a brief look at four ways to quickly rank these improvement opportunities: forced ranking model, risk matrix, paired comparisons model, and weighted criteria model. Chapter 5 presents tools that we can use to investigate a specific process for improvement possibilities: My own variations on a problem-solving model, fishbone diagrams, force field analysis, Pareto diagrams, process charts, and root cause analysis.

Chapter 6 discusses the life cycle of projects that are focused on improvement initiatives. These are projects just like any other projects we have been managing for years but with some notable differences. They are small, they are frequently killed, their scope can change radically, and they give life to many new improvement projects. I will lay out the life cycle of these projects along with the documentation specific to them.

Chapter 7 presents a case study that I call B. Stoveburden Trucking Company. It is a disguised version of an improvement program from one of my clients. Not all of their adventure is reported here. I have condensed it to contain the most significant analyses and findings. It shows by way of example how all of the tools introduced in Chapters 5 and 6 can be used.

Chapter 8 is the closing chapter and presents two important thoughts with which I wish to leave you. The first deals with implementation challenges. Improving the effectiveness of your project management process is no small task. It embodies value statements, cultural fit, and sponsorship. The second is implementation strategies. There are at least three ways to approach project management process improvement: through major project initiatives, through a long-term program, and through a slow but steady pace.

Hopefully you have a clear picture of my agenda for this book. You will find my presentation style to be informal and brief. I do not want to burden you with more reading than is necessary. The purpose here is to give you a handy guide and reference for your improvement initiatives. You have my best wishes for a productive and rewarding reading and learning experience.

1

Introduction to the Process Improvement Life Cycle

Designing, documenting, and implementing a project management methodology is a major undertaking. It is met with several obstacles, including:

- Cultural and organizational barriers to change;
- Replacing existing project management habits;
- Rugged individualism of technical professionals.

An organization will never reach the point where it is safe to say that all three of these obstacles have been neutralized. In fact, these obstacles will continuously plague projects for as long as there are projects to be plagued. Until very recently most organizations have not paid enough attention to these obstacles, which ushered in the beginning of the end for many of them. Once the methodology is introduced to the organization, however, its founding fathers claim success and the process of implementing a project management methodology officially ends. To be specific, reaching this point is just the end of the beginning. The strategy for the middle game involves direct confrontation with the above obstacles.

Project failure rates are expected to decrease as a result of using the new methodology, but they do not. Project teams are supposed to use the new methodology, but they are not. The problem is often serious enough to require commissioning a project to find a solution and implement it. As organizations come to the conclusion that the time and cost expended to reach the point of full project management methodology implementation is an investment in the future, attitudes change. These organizations turn their attention towards protecting their investment, and so, programs at improving project management performance are sought. This chapter introduces the solution at the process level. Later chapters expound on that solution.

1.1 The Importance of Process Improvement

The amount of effort put into the design and implementation of a process does not really matter; there is always room for improvement. Nowhere is this more obvious than in the technical professions. As new technologies emerge, new ways of doing things arise and we must change or die. In other words, an organization simply cannot stand still and expect project management to continue to function at expected levels of effectiveness. It must continuously improve processes or they will fall into misuse or no use at all.

1.1.1 Stand Still and Go Backwards

A professional athlete is good in part because of countless hours of practice. A professional athlete stays good because of countless hours of practice. The analogy to business is that a competitive business is good by constantly improving what it does. A competitive business stays good by constantly improving what it does. The analogy to project management is that to be good at project management requires the committed and organized effort of the entire organization, and to stay good at project management requires that continual effort.

To ignore this warning is to sentence your business to a slippery slope that sooner or later will lead to failure. The first signs that your organization is on that slippery slope is a complacent attitude that creeps into everyday business life and project life that everything is fine, and that you have arrived at an exemplary practice of project management. In the final analysis, all organizations should look for ways to improve the quality and maturity of their project management processes. Before doing that it would be informative to look at the sources of problems regarding the process and its practice. For that information we turn to the Standish Group for their insights into information technology (IT) project success and failure. Once we have digested that information we will be in a better position to put together an action plan.

1.1.2 Standish Group Chaos Report

Beginning in 1994 the Standish Group, an independent IT research organization, published the results of extensive interviews with IT executives regarding reasons why IT projects succeed or fail. Their report, "CHAOS Chronicles,

Version 2.0" [1], lists the following 10 reasons for project success in the order of importance:

- Executive support;
- User involvement;
- Experienced project manager;
- Clear business objectives;
- Minimized scope;
- Standard software infrastructure;
- Firm basic requirements;
- Formal methodology;
- Reliable estimates;
- Skilled staff.

Seven of the 10 reasons relate to process. The remaining three (executive support, experienced project manager, and skilled staff) relate to the project, or more specifically to the alignment of the project manager and the team members to the project as well as the alignment of the project to the organization's goals and objectives. For a detailed discussion of how that alignment is measured and acted upon, consult my companion book *Building Effective Project Teams* [2].

It will be helpful to review each of these 10 reasons especially as they relate to the quality of the project management process. The discussion below will focus on the practices and processes that must be part of a quality project management methodology. This will lay the foundation for our assessment of project management maturity later in the book.

1.1.2.1 Executive Support

Since this is the single-most important reason for project success, its absence is the main reason why projects fail. While some may think it is simple to get executive support, many would agree that it is not easily maintained. Changes in executive leadership, changing political scenes, and changing business priorities can easily result in the loss of that support. Because of this fragility, the project manager must be held to standards and practices that preserve rather than alienate the executive sponsor. Those standards and practices will be part of the communications management program that makes up the project management methodology.

Executive management must have a stake in the outcome of the project. A well-devised project plan, along with project team commitment, will go a long way in gaining executive management buy-in. And if the executive becomes the leading spokesperson for the project, it is a sure sign of management buy-in. The executive should be a visionary, setting the agenda, arranging funding, articulating objectives, and also be the champion and minesweeper, securing necessary resources and taking total ownership of the project. The executive should not be the project manager, or the function representative, or Santa Claus, or the technical officer.

Executive support must go beyond their pet projects and extend to the project management methodology itself. Their endorsement of the methodology as the efficient and effective way to manage projects must be visible. They have to walk the talk!

1.1.2.2 User Involvement

The best way to assure user help and support when it is needed is to keep the user meaningfully involved in the project throughout its life cycle. That begins with functional specification, continues through planning and execution of the project, extends into change management and problem solving, and culminates with a well-defined acceptance criterion.

Even though a project is on time and within budget, it can fail if users' expectations are not met. The project team must understand the users' business and their needs and effectively communicate with them. The users need to provide constant information and feedback and can do so through formal (meetings) and informal methods established by the project team. There must be mutual respect between users and the project team. The "correct" users must be involved early and often in the project life cycle and they need to own the project. A function representative is the "voice" for all user departments and serves as the subject matter expert. There are many opportunities in the project management life cycle to meaningfully involve the user, your client. The extent to which your defined processes include the client is an indicator of project success.

1.1.2.3 Experienced Project Manager

Availability is not a skill! To appoint someone project manager simply because they are available is not a good management decision. Such behavior is probably rare but it should make us stop and think about exactly how we do appoint a person project manager. That decision should be based on a number of factors, which can be summarized in one observation: How well does their skill and competency profile match the characteristics of the project? The question cannot be answered unless we have a way of profiling projects, a way of profiling the skills and competencies of the project manager, and a way to measure the alignment of the two profiles. The interested reader can consult *Building Effective Project Teams* [2] for a metric that measures the alignment of the two profiles.

Business and technical knowledge, judgment, negotiation, organization, and good written and oral communication skills are desirable traits for a project manager. The ability to communicate with all the stakeholders and technical teams is necessary. Additionally, planning, tracking activities, tasks, and changes, or replanning to arrive at a goal are other skills a project manager should maintain. A project manager should decide what features and functions are part of the project, orchestrate all resources, focus on the goal and minimize diversions, and establish accountability, responsibility, and authenticity. A project manager should not be the executive sponsor, user, or functional representative, and should not overpromise or be a control freak.

1.1.2.4 Clear Business Objectives

Project management methodology must have a formal process for establishing clear business objectives. If you do not know where you are going, how will you know what to do and how will you know when you get there? Projects that start out without having this information are in trouble unless the methodology has a way of compensating for that lack of information. Traditional project management approaches do not have a way of compensating; the newer adaptive and iterative approaches do. Since change is almost certain, the project management methodology must have a way of maintaining the objectives as they change and a way of adapting the project plan to those changes.

Everyone associated with a project must share the same vision. The vision must be clear, concise, and comprehensible. The goal(s) of the project must be known and enthusiastically supported by all. And goals must have measurable success factors. The project's business objectives must map to the corporate vision. This ensures that those associated with a project know and understand the objectives, where they fit in, and how the project goals contribute to the corporate vision.

Despite all of the effort devoted to clearly defining project goals and objectives, these are not static and will change. You may have been very successful in working with your client to achieve that clarity, but it may not be long lasting. Business conditions will change, markets adjust to the economy and to new competition, and competitors will change. All of these factors lead to scope change in your project and place your project at risk. That means your project management process must have a solid change management process that is integrated into other business processes.

1.1.2.5 Minimized Scope

The trade-off here is that longer projects will incur more change and risk and less so for shorter projects. Change in scope brings about a change in the project plan and the increased risk that work completed earlier may no longer be of value. That means wasted dollars and wasted time. A large project can be decomposed into several interdependent smaller projects. Each smaller project should be justified based on the specific deliverables and business outcomes that will be produced. The extent to which the project management methodology considers this approach and includes processes for decomposition is a measure of its quality and maturity.

Major milestones in a project form the boundaries from one phase to the next. Adding some smaller milestones and monitoring their attainment is one of the keys to project success. The five key elements to effectively using project milestones are planning, top-down design, time boxing, tools, and management by objectives/accountability. Proper planning prevents problems. Start with a high level view then figure out the details. Time boxing involves set deadlines and a fixed amount of time. Using automated tools and templates can speed projects up. Milestones must be defined, understood, measurable, and quantifiable. And each should have an assigned owner.

1.1.2.6 Standard Software Infrastructure

This factor speaks to the stability of the infrastructure over which your project work will be done. If that infrastructure is in flux, your project plan is at risk for radical change. That risk opens the possibility of missed deadlines, use of the wrong human resources, team members with the wrong skills, inability to meet the client's requirements, and a host of related impacts.

It is vital to use a language that is understood by all parties involved in a project. Infrastructure is defined as the underlying foundation or basic framework (as of a system or organization). Defining, understanding, and engaging standard business processes is fundamental to any company, and that includes ensuring a standard business infrastructure throughout the enterprise environment. A standard technology infrastructure can facilitate the placement of new kinds of technology to support business initiatives. Selecting a robust and scalable infrastructure will enable businesses to profit and expand by harnessing the capabilities and promise of truly global electronic commerce.

1.1.2.7 Firm Basic Requirements

This is a no-brainer. Much of the discussion surrounding clear business objectives applies here. By way of analogy, you cannot start out on a journey unless you have some idea of where you intend to go. The better you can define that journey the more effective will be your initial choices of direction. The better you understand the client's basic requirements, the better your plan will be for delivering an effective solution to meet all of their requirements.

Requirements management is the ongoing process of identifying, documenting, communication, tracking, and managing project requirements as well as changes to those requirements. The earlier an error is detected, the less costly it is to fix. A concise definition of the project vision should be written in business terms. Buy-in from the users and executives are paramount to project viability. Continuous reevaluation must occur. Identify all key stakeholders and include them in the requirements definition. Identify and document all risks and formulate a plan to minimize them. Develop a clear statement of the business case. Define the project metrics, measurements, and milestones.

1.1.2.8 Formal Methodology

Project management methodologies that can be repeated are valuable to the organization. Repeatability creates standards, best practices, skill development, and a host of other benefits to the organization. Project management methodologies that are adaptable rather than rigid are valuable to the organization as well. The extent to which a project management methodology is standardized, documented, accepted and practiced, integrated into the business equation, and improved upon is a measure of its quality and maturity.

The project management office (PMO) is part of the infrastructure that will help an organization align business and technical goals and increase the odds of project execution in organizations. It is a dedicated section of the organization that focuses on various aspects of project management and methodology. PMOs help to gain better control over processes and project outcomes, bring consistency to their implementations, standardize operations, control resource allocation, and handle customer interfacing. PMO staff members have project management experience and excellent communication skills.

1.1.2.9 Reliable Estimates

Historical estimated versus actual costs and durations are your best tools for producing new estimates of cost and time. The availability and maintenance of this historical information is a sign of the maturity of the project management process.

Reliable estimates can only come from honest and frank assessments. It is important to create realistic written specifications, prioritize needs, and work toward smaller milestones at frequent intervals. Managing change is another requirement in setting realistic expectations. A misalignment between expectations and deliverables often occurs if change is not managed.

1.1.2.10 Skilled Staff

There are two factors to consider here. The first is the skills inventory present in the staff and the extent to which it matches the demand for skills in the organization. The second is the extent to which the skills of the project team match the skill requirements of the project to which they have been assigned.

Skilled staff is your most valuable asset. The five key elements to ensure competency are:

1. Identifying required competencies;
2. Providing a quality, relevant, and continuous training program;

3. Recruiting both internally and externally;
4. "Incentivizing" the staff;
5. Ensuring they are project-focused.

Building and maintaining a team involves collective participation from the entire team. Communication within a team is vital to a project's success.

This book will focus on these 10 reasons that relate to the effectiveness and maturity of the project management process. More background needs to be provided before we can meaningfully discuss these reasons. Specifically, we need to describe the processes that comprise a typical project management methodology and then relate those processes to these 10 reasons. That will be the topic of Chapter 2.

1.1.3 Balancing People, Project Management Processes, and Technology

Each of the 10 critical success factors tells us a great deal about the characteristics of effective project management processes, but they do not tell the whole story. In addition to the processes, an effective project management environment is also made up of people and the technology to support the processes and the people. The 10 critical success factors tell us that. Four are related to people, seven are related to process, and three are related to technology.

The triad formed by people, project management processes, and technology forms a system that must be in balance if projects are to have any reasonable chance of succeeding [3]. Figure 1.1 displays that triad.

The figure shows several examples of how the three components can be related to one another. These are illustrated by data points A, B, C, and D. The closer the data point is to a vertex, the more developed or stronger that component is to the mix. Data point A is located at the center of gravity of the triangle and represents a system in balance. The People dimension shows a staff whose skill profile and experience level is in balance with the needs of the organization. The Project management processes dimension shows that the organization has sufficiently developed and understood project management processes to meet the needs of the organization. The Technology dimension shows that the organization has deployed the appropriate level of technology to support the project management processes that are in place and the people who use those processes.

Data point B shows an organization that is tilted toward the Project management processes and Technology dimensions. This organization will have sophisticated project management processes in place and the necessary supporting technology. They will not be effective, however, because they have not adequately prepared their people with the training and skills to effectively utilize

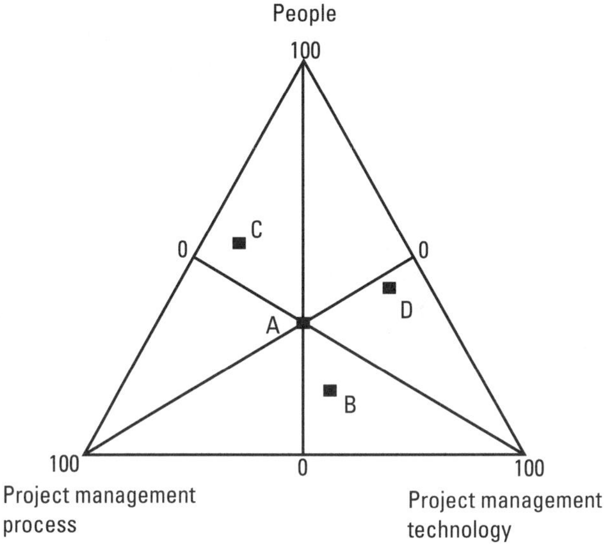

Figure 1.1 The triad of people, project management processes, and technology.

the infrastructure. Furthermore, while the project management processes themselves may be entirely appropriate, the people may not be using them. This illustrates that a gap exists between one or more project management processes and the practice of those processes. This situation will be a major topic of discussion in Chapters 3, 6, and 7.

Data point C shows an organization that is tilted toward the People and Project management processes dimensions. This clearly shows a failure on the part of the IT function to support the project management function and on the part of the project managers to proactively go after technologies to support their project work.

Data point D shows an organization that is tilted toward the People and Technology dimensions. This is a fairly common occurrence. These organizations are technology rich and have hired smart people, but without the processes to support their business activities turnover will be high and business will suffer as a result.

1.1.4 Process Improvement Versus Practice Improvement

The effectiveness of project management in an organization is measured by two separate but related maturity variables.

The first is the current maturity level of the process itself. The assessment of this maturity is based on an evaluation of the standardized and documented methodology for managing projects in place in the organization. The assessed maturity level will be a point estimate. For any one of the many processes that make up a project management methodology, their scope and impact on the organization will increase over time. This will happen as shortcomings in the initial version are discovered and fixed. Increases in scope will occur as the project management processes become more integrated into related business processes.

The second variable is the maturity level of the practice of the process as evidenced by ongoing projects. This assessment will be done on projects recently completed and will be repeated at set intervals (such as quarterly). The assessed maturity level will be a distribution of values—one for each assessed project.

There are two questions that can be answered from this data. The first is whether the current process maturity has reached the level established by the organization as the target level. If not, part of the effectiveness improvement will, of course, be to put initiatives in place to reach that target level.

The second question is whether the current practice maturity is consistent with the process maturity. There are three situations that will be important to us as we examine current effectiveness. They are introduced in the sections below and will be further discussed in Chapters 3 and 4.

1.1.4.1 Process Maturity Exceeds Practice Maturity

This situation will occur when the organization has a process in place that has not yet been fully integrated into practice. There can be many reasons for this:

- The process may not have been successfully deployed into the organization.
- The process may not be sufficiently documented.
- The process may not be appropriately defined and has therefore been dismissed as not useful or misused by project teams.
- Process training may not be effective.

Project reviews should be able to get to the root cause of this performance gap. Chapter 5 will take up this situation and show how that root cause can be determined by way of an example.

1.1.4.2 Process Maturity Equals Practice Maturity

This suggests a healthy alignment between the process and how it is being practiced. Keep in mind that the practice maturity data will have been collected over several projects and presented as a distribution. Not all projects will have verified a practice equal to the process maturity level. Rather, some will be above and some will be below that level. In the aggregate, however, a reasonable person would conclude that the practice maturity mirrors the process maturity. The evidence to support this conclusion would be a distribution of practice maturity values around the process maturity. The smaller the variance of that distribution, the more aligned are the practice and process maturity.

1.1.4.3 Process Maturity Less Than Practice Maturity

This would seem to be an anomaly but it really is not. It is entirely possible that projects will display a practice maturity that is above the process maturity for at least the following reasons:

- The process maturity level has not been documented and deployed and several project managers have taken it upon themselves to practice at a maturity level that is called for.
- One or more members of project teams have brought practices with them from other organizations that exceed the process maturity level on one or more processes.

These situations are likely to arise when the process maturity levels are lower compared to what the organization would expect them to be. As the process maturity levels increase, this phenomenon is less likely to occur. But while it does, there are best practices and lessons learned that can be gleaned from these projects. In fact, these best practices and lessons learned can become the input to improvement initiatives for the process itself.

1.2 Typical Project Improvement Practices

Initial attempts at improving the practice of project management emanated from experiences gained on individual projects. Three efforts can be highlighted: project reviews, best practices, and lesson learned.

1.2.1 Project Reviews

At major project milestones, review meetings will often be held. While the primary purpose of these is to assess the general status and health of an ongoing project and offer get well plans as appropriate, they also present opportunities for discussing the use of documented project management processes and other practices.

These discussions can be very illuminating. If the project team has taken variance from an established process, the reason for the variance should be brought to light. The variance may be due to the fact that the process as defined does not meet the specific needs of the project. The variance might also be due to the team having an alternative process that it believes is better than the established process. The variance may be due to ignorance on the part of the project team. In all of these cases there is valuable intelligence to be gathered. The intelligence can lead to improvements in process as well as practice of process.

1.2.2 Best Practices

By attending conferences, reading, and talking with project managers at other companies or those recently hired into their company, project managers pick up good ideas, practices, and processes. Through their project assignments they have an opportunity to use what they have learned. This will certainly help them with their current assignment. If there is a process in place to transfer that knowledge to the organization in a useful form, all future projects can benefit.

1.2.3 Lessons Learned

Through a postimplementation audit, project managers are supposed to pass on what they have learned about process and practice to other project teams that follow them. They should have learned new approaches and new strategies for improving the management of their projects that will be of value to others to follow. They might also discover approaches that simply do not work. That information should also be passed forward for use by other projects. In theory, all of this sounds good, but in practice, it just does not happen. The reasons for this are discussed in the next section.

1.3 Definition of the Process Improvement Life Cycle

While the above improvement practices have been in use for a number of years, the results have not met expectations. There are several reasons for this:

- No interproject sharing of ideas and suggestions.
- Each project is unique.
- Experiences do not travel well from project to project.
- The process is too informal to work.

Senior level managers expect that project managers will share their experiences so that others might learn and thereby improve their project management practices. This does not happen to any measurable extent, and the reason most often given is that "my project is so different from yours that your idea simply doesn't apply." Project managers are often hesitant to take ideas and suggestions from other project managers because it may be a sign of weakness or incompetence. The problem with these reasons is that they are all the result of the informality involved. Depending on an informal spreading of ideas and suggestions simply will not work. Having a formal process in place to receive ideas and suggestions and filter them for general use might solve the problem somewhat. If the organization expects improvement in its project management practices and processes, it will require a formal and planned approach. That approach is the process improvement life cycle, which consists of four major steps as shown in Figure 1.2.

1.3.1 Where Are You?

The first step is to determine where you are. I am not talking about a physical location but rather about a state of being. The question is really asking for some statement about process maturity. To answer that question we must have some basis for measurement of process maturity. Chapter 2 develops a survey to measure that. Chapter 3 develops a metric from the resulting survey data. The answer to the question "Where are you?" is given by the value of the process metric. We will call this the *baseline*. Over time, and as the process and its practices improve, comparisons against that baseline will generate a trend line showing changes

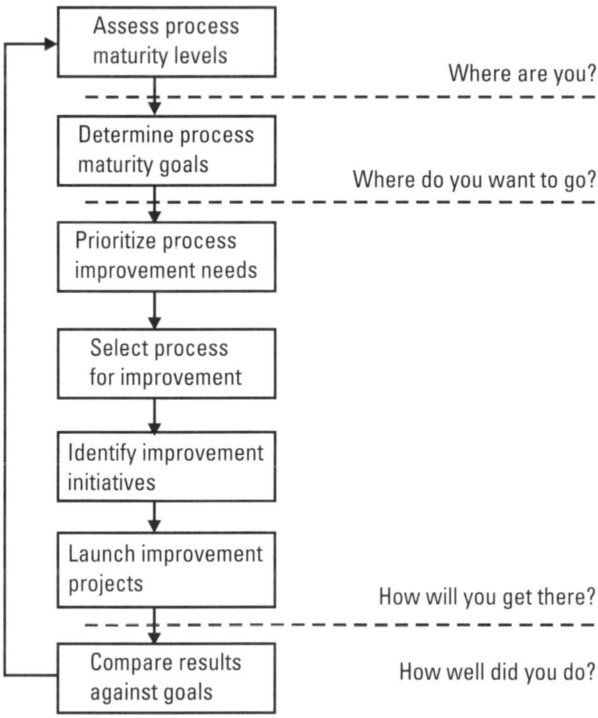

Figure 1.2 The process improvement life cycle.

with respect to the baseline. In other words, the answer to the where are you question changes. If the organization has a quality improvement program underway, the trend line will be a tool for measuring progress as the baseline converges on the target maturity level.

1.3.2 Where Do You Want To Be?

Once you have determined where you are, the next step is to decide where you would like to be. What goal will you set for your process improvement efforts? The goal should be expressed against that same baseline and in terms of the metric used to establish your current state. That goal can be very short term and associated with a specific improvement opportunity or a long-term programmatic goal associated with a final end state.

1.3.3 How Will You Get There?

At this stage you know where you are and you know where you want to be and both are expressed in terms of the same metric. The difference between the two is called your *process maturity gap*. To answer this question we have to define the pathway that connects the two end points that define the gap. That pathway is a plan to move the people, project management process, and technology that define the current state to another combination of people, project management process, and technology that define the end state. There will be cases where that pathway is clearly defined and others where that pathway is only vaguely defined. It all depends on the complexity of the relationship between people, project management process, and technology along that path. Each situation will require a different project approach. These will be discussed in Chapter 6.

1.3.4 How Well Did You Do?

An improvement program has been put in place to narrow the gap. That improvement program may consist of a single project or several projects. The projects could be totally independent of one another or related in some manner. In most cases the projects are sequential. The results of the first project are assessed with respect to the defined goal. If the goal was reached, other improvement programs may be initiated for other processes. If the goal has not been reached, a root cause analysis may be conducted and additional projects commissioned as a result. In any case, there is a final assessment of the new baseline. The life cycle then repeats itself continuously.

1.4 Who Is Responsible for Process Improvement?

Without any reservations, the answer to this question is the project support office (PSO). There must be a single point of responsibility for several reasons, such as:

- Establishing a standard processes;
- Managing best practices and lessons learned;
- Managing performance data against standard processes;
- Continuously improving the project management process.

1.4.1 Establishing a Standard Process

Managing and improving project management process standards are not possible unless there is some coordinating unit that has been given responsibility for the development, deployment, and maintenance of the standards. Developing standard project management processes should be a collaborative effort involving PSO staff and a representative task force of project managers. That collaboration will go a long way towards contributing to ownership and a successful implementation. The standard should not be "one size fits all." Rather, it should offer options to the project manager based on the characteristics of the project.

1.4.2 Managing Best Practices and Lessons Learned

This responsibility extends from putting processes in place to identifying and collecting best practices and lessons learned to cataloging and creating access to them, to distributing them throughout the organization. None of these is a trivial responsibility. Experience has shown them to be very difficult. Best practices must include activities that can be isolated from any specific inputs or outputs. In other words, they must become robust if other projects are to use them. This means that someone has to take the responsibility to look beyond the best practice or lessons learned as submitted and extract their real essence. This will allow the best practices and lessons learned to be used by others regardless of the dependent processes they might be using in conjunction with the so-called best practices. Lessons learned are perhaps a bit simpler to deal with. Many, if not most, of them can be recorded in if-then types statements. If this is the situation you have encountered, then this is the suggested action.

1.4.3 Managing Performance Data Against Standard Processes

The need here is for an unbiased party to conduct project reviews that will result in consistent data being generated across projects so that meaningful analyses can be made and conclusions drawn. Project reviews are usually conducted by a panel of project managers at designated milestone events in the life of a project. The purpose of a project review is to assess how well the project plan and the documented project management processes are being followed. The intent is not only to assess project status against the plan but also to help the project manager. In some cases the project manager is able to share a tool or technique they learned or developed and contribute it to the store of knowledge on project management for the organization. These best practices can be useful to other project teams.

1.4.4 Continuously Improving the Project Management Process

Again, this must be coordinated from a single point of reference. Almost without exception this should be a PSO, which will have the responsibility of developing and implementing the project management methodology for the organization and will also be responsible for monitoring compliance with it. Through project reviews and other inputs on project performance, the PSO will be able to see areas where the process can be improved and areas where the team may need some additional training or consulting support.

1.5 Effectively Dealing with the Obstacles

The introduction to this chapter listed three obstacles to designing, documenting, and implementing a project management methodology:

1. Cultural and organizational barriers to change;
2. Replacing existing project management habits;
3. Rugged individualism of technical professionals.

Let us now take a brief look at the strategies and tactics for neutralizing these obstacles.

People grow comfortable with processes and procedures and any attempt to change them is met with resistance. That resistance stems from a fear of the unknown, a fear that the change may invade their space, or a fear that they will lose power or prestige. Regardless of the impact of the change, these perceptions must be treated as reality. If the patterns of behavior that the individual follows are the result of processes that they personally developed to meet their needs, it will be even more difficult to replace their behavior with new processes. As a group, engineers often present even a greater challenge. For too many years engineers lived under the myth of "not invented here." Engineers are perfectly capable problem solvers and process developers. They behave as though they are self-sufficient and can build whatever they need to get their jobs done.

Knowing these patterns of behavior ahead of time should warn us to thoughtfully plan a strategy and the necessary counter measures before we attempt to introduce change. That plan begins at the time management decides to move ahead with the change. I contend that these obstacles can be effectively addressed if you adhere to the following principles:

- The project team should consist of credible and influential representatives of each constituency that will affect or be affected by the new processes.
- The project plan should contain a component that designs the new process at the conceptual level.
- The project plan should contain a significant component that will seek out best practices inside and outside the organization.
- The project plan must contain several review and open meeting sessions that investigate draft versions of the new process.
- No one should be excluded from offering their opinions and being listened to.
- Visible response and closure to every suggestion must be made.

1.6 Points to Remember

The following is a list of important points to remember from this chapter:

- Once a project management methodology has been introduced to the organization, attention must shift to strategies for implementation.
- If the project management methodology is not continuously improved, it will fall into misuse or no use at all.
- To be good at project management requires the committed and organized effort of the entire organization, and to stay good requires that continual effort.
- Seven of the 10 reasons for project success are related to the project management process itself.
- The triad formed by people, project management process, and technology forms a system that must be in balance if projects are to have any chance of success.
- The effectiveness of project management in an organization is measured by the maturity level of the process (PD) and the maturity level of the practice of the process (PP).

- The process improvement life cycle consists of answering four questions:

    1. Where are you?
    2. Where do you want to go?
    3. How will you get there?
    4. How well did you do?

- A PMO will be needed if an organization is serious about process improvement.
- Putting a project management methodology in place and continuously improving it will face barriers to change, the need for individuals to change their habits, and the rugged individualism of project managers.

References

[1] Standish Group, "Chaos Chronicles, Version 2.0," West Yarmouth, MA, 2001.

[2] Wysocki, R. K., *Building Effective Project Teams*, New York: John Wiley & Sons, Inc., 2002.

[3] Wysocki, R. K., and R. L. DeMichiell, *Managing Information Across the Enterprise*, New York: John Wiley & Sons, Inc., 1997.

2

Overview of the Project Management Maturity Model

2.1 The Software Engineering Institute Capability Maturity Model®

Beginning as early as 1986 the Software Engineering Institute (SEI), which is affiliated with Carnegie Mellon University, began developing a process maturity framework for software development [1]. With financial support from the Department of Defense this early effort resulted in the publication of the Capability Maturity Model® (CMM®) [2] in 1991.

This is a lengthy foundation chapter in which the detailed description of the five-level maturity model is presented and applied to each of the 39 processes that define the project management body of knowledge (PMBOK). These descriptions provide the content for the survey that will be used to measure process and practice maturity. Maturity assessment will be the basis for a continuous improvement program for project management processes.

2.1.1 Purpose

The purpose of the CMM® is to provide organizations with a guide for establishing process improvement programs for software development. The guide can be used both as a foundation for establishing tools and as input to creating a maturity questionnaire for process improvement.

2.1.2 Structure

The CMM® defines five levels of maturity: initial, repeatable, defined, managed, and optimizing, which are briefly described below.

2.1.2.1 Initial

The process is ad hoc. There may be a few defined processes. Some software engineers bring tools and templates they may have learned elsewhere but otherwise successful software development is largely dependent upon heroic efforts.

2.1.2.2 Repeatable

Processes are established and put in place for use across software development projects. Process use is recommended but not required. For some large or critical mission projects the use of these standard processes is often required.

2.1.2.3 Defined

Processes are standardized and documented. There is a standard software development process that all projects must use. Training and support are available through a PSO.

2.1.2.4 Managed

Project progress against plan is monitored, reported, and controlled. Decisions regarding software development projects are made with reference to organizational considerations. Project management decisions are integrated into other business processes.

2.1.2.5 Optimizing

Project performance is fed back into the process itself to enable a continuous quality improvement program. Best practices and lessons learned are input to the improvement program.

2.1.3 Application

It turns out that the CMM® is quite robust and has application beyond software engineering, for which it was originally developed. There are two areas of application that it has spawned. They are the People Capability Maturity Model® (P-CMM) and the Project Management Maturity Model (PMMM). They are described below.

2.1.3.1 People Capability Maturity Model®

The P-CMM® is a five-level model patterned after the five levels of the CMM®. Except for level 1, each level is comprised of a number of process areas as listed in Table 2.1 and described below.

Table 2.1
Levels of the People Capability Maturity Model®

Level 1	Initial	No processes defined at this level
Level 2	Managed	Staffing Communication and coordination Work environment Performance management Training and development Compensation
Level 3	Defined	Competency analysis Workforce planning Competency development Career development Competency-based practices Workgroup development Participatory culture
Level 4	Predictable	Competency integration Empowered workgroups Competency-based assets Quantitative performance management Organizational capability management Mentoring
Level 5	Optimizing	Continuous capability improvement Organizational performance alignment Continuous workforce innovation

Staffing (2)

Level 2 staffing activities include the assignment of qualified individuals to tasks based on the degree to which their skills align with the requirements of the task to which they are being assigned. Obviously then there must be a formal selection process in place to assure a fair and equitable evaluation and assignment of each individual. In addition to the initial assignment decision, the selection process should also include procedures for moving people to new positions within the same or different assignments.

Communication and Coordination (2)

The focus of this level 2 process is open and timely communications across organizational units. This includes coordination of activities across units that are dependent upon one another for the effective and efficient completion of these activities.

Work Environment (2)

This process includes both the provision of both the resources needed to complete assigned tasks as well as the environment in which those tasks are undertaken. The monitoring and provision of this environment is a management responsibility.

Performance Management (2)

The focus of this level 2 process is the establishment of ways to measure performance of the individual and the processes they use to do their work. The ultimate management goal is the improvement of both forms of performance.

Training and Development (2)

This process involves providing the training needed to close any gaps that exist between the skills possessed by the team members and the skills required of the team members in order to meet their assigned responsibilities.

Compensation (2)

Compensation requires an organization-level strategy to assure fair and equitable compensation for an individual's contribution and value to the organization. Having such a strategy in place gives some impetus to skill development and better alignment of the individual to the needs of the team.

Competency Analysis (3)

This level 3 process identifies the knowledge, skills, and abilities needed to meet expectations for the organization's business activities. The process includes a provision for measuring, storing, and maintaining the individual's knowledge, skills, and abilities so that the organization's capabilities in each competency area can be accurately assessed.

Workforce Planning (3)

Based on the above competency analysis and the demand for workforce skills to meet the organization's current and future needs, a plan is developed to meet those needs. The plan assures that the required workforce skill profile will be available when needed.

Competency Development (3)

Competency development flows from the previous two processes. Competency analysis identifies the current skill and competency profile of the workforce. Workforce planning defines the current and future skill and competency needs of the workforce. Competency development is the planning of the training and development needs and the execution of that plan.

Career Development (3)

This process focuses on the development of individual plans to facilitate the definition and achievement of career goals for each individual. The process includes a monitoring capability as well. The plan identifies the career progression that will lead to the career goal and a method of updating that plan.

Competency-Based Practices (3)

This is an integrative process. It coordinates the output of the previous processes at the managed level to assure that workforce activities support the attainment of organizational goals.

Workgroup Development (3)

In the context of this book this process applies to team formation and deployment. Teams are formed based on the collective skills and competencies needed to successfully meet the client's requirements.

Participatory Culture (3)

This process forges the members into a high-performance team. Such team activities as decision making and problem solving result from the creation of this participatory culture.

Competency Integration (4)

This level 4 process integrates the processes that were defined at level 3. It recognizes and establishes the interdependencies that exist between skills and competencies as evidenced within the work activities that utilize these competencies and skills.

Empowered Workgroups (4)

This process involves delegating responsibility and authority of the workgroup to carry out the tasks in their work activity. The workgroup becomes an entity that management interfaces with through training and other management interface activities. Empowered workgroups have total responsibility for the success of their work activities. This includes recruiting, selection, performance monitoring and management, training, development, and compensation.

Competency-Based Assets (4)

This process focuses on sharing best practices among work groups. It encompasses not only the collection and storing of best practices but also mechanisms for sharing those best practices. Best practices include templates, tools, and other artifacts that are developed during the course of work activities that will have value to other work groups.

Quantitative Performance Management (4)

The workgroup prioritizes the competency-based processes that relate to the successful achievement of their workgroup objectives. Performance baselines are established and workgroup performance is measured against those baselines. Any performance variances that are significant lead to corrective actions.

Organizational Capability Management (4)

The focus of this process is on the quantification and management of each workforce's capability in the performance of the processes that are critical to its unit's objectives. These capabilities are measured, analyzed, and managed. The results are used to adjust organizational competencies to improve results.

Mentoring (4)

The focus of mentoring is the transference of competencies and skills to the workforce from the more experienced to the less experienced. Mentoring focuses on the knowledge, skills, and process abilities as well as on deployment. A program of mentor selection and training is implemented.

Continuous Capability Improvement (5)

This level 5 process focuses on establishing the infrastructure upon which individuals will have the foundation to improve their ability to perform skill and competency-based processes. Training at the individual level and integration of individual skills and competencies are undertaken to improve the performance of work groups.

Organizational Performance Alignment (5)

This process deals with organizational efforts to align the skills and competencies of individuals and work groups to meet organizational performance and business goals. The alignment efforts involve analyses of competency capabilities with competency requirements.

Continuous Workforce Innovation (5)

This process focuses on the continuous improvement of workforce practices by identifying and integrating individual and workgroup practices across the organization. It is basically a best practices discovery and integration effort.

2.1.3.2 The Project Management Maturity Model

The other major adaptation of the CMM® is to project management maturity. The PMMM adapts the five-level CMM® to the process and practice of project management. Since the PMMM is the major focus of this book, I have devoted an entire section to the discussion of its structure. As we will see, it provides a validated foundation for the improvement of both the process of project management and practice of using and adopting that process. It is my belief that the organization can have an exemplary project management process in place, but if the adoption of that process is lacking, the accomplishments are reflected only on paper and not in performance. The major thrust of this book is to improve the process and at the same time assure that the gap between the process and the practice of the process is managed. Maturity level targets will be established for both process and practice. The improvement program will be designed to realize those targets.

2.2 The Project Management Maturity Model

The CMM® first refers to project management at level 2, where the focus is on repeatability, and hence begins the definition of standards for project management. The PMMM takes these standards to the next level of development by defining a separate model for the process and practice of project management. This model parallels the CMM® as described below.

2.2.1 Level 1: Initial Process

This level can almost be renamed the "Do it yourself" or "Do it your own way" level. There are no standards and project management processes are ad hoc. There may be an awareness of practices followed by other projects, but their use is entirely at the discretion of the project manager. This does not mean that projects will necessarily fail or be subject to poor management. In fact, for a given project the practice of project management is largely dependent upon the process knowledge possessed and practiced by the team members. It may be very poor. On the other hand, there may be excellent practices but they are known only within the team itself. There is no organized way to share these best practices outside the team. Because of the ad hoc nature of this type of project management, these best practices may not travel very well. That is, they may not be very useful to other teams who are practicing project management their own way. Management may well be aware that there are project management processes and standards but there is no evidence that any movements have been made to establish them in this organization.

There are a few characteristics of level 1 maturity that apply regardless of the process:

- There is no defined and documented process in place.
- Project managers and teams act in an ad hoc manner when process activities are needed.
- Processes and practices may be taken from prior experiences or knowledge possessed by one of the team members.

2.2.2 Level 2: Structured Process

At level 2 a number of project management processes exist within the organization and most are documented. However, there is no requirement that projects use these practices. Project teams will use these processes when it suits their needs even though management encourages their use. Status reporting of project progress against plans is ad hoc and not consistent across projects.

There are a few characteristics of level 2 maturity that apply regardless of the process:

- There are defined and documented processes in place for team use.
- Project managers and teams use the defined processes at their discretion.
- Critical mission projects are often required to use the documented processes.

2.2.3 Level 3: Institutionalized Process

Level 3 is differentiated from level 2 by the adoption of a standard that is required by all project teams. The standard allows for the adaptation of the processes and practices to the particular characteristics of the project. There is no "one size fits all" mentality.

There are a few characteristics of Level 3 maturity that apply regardless of the process:

- There is a comprehensive defined and documented process in place that is used by all projects.
- There is support available to teams needing help with the standard processes.
- There is a monitoring and control function in place to assure compliance with standard processes.

2.2.4 Level 4: Managed Process

Project management and other corporate management systems are integrated. There are metrics in place to compare performance across the project portfolio. Senior management understands its role in managing the project portfolio.

There are a few characteristics of level 4 maturity that apply regardless of the process:

- The process is integrated into other business processes and practices.
- Management decisions on individual projects have an organizational perspective.
- Lessons learned and best practices are captured and made available to other projects.

2.2.5 Level 5: Optimizing Process

At level 5 the focus is on improvement of the project management process. To that end, processes are in place to identify and take action on performance issues related to process, and to incorporate best practices and lessons learned as feedback to project management process improvement.

There are a few characteristics of level 5 maturity that apply regardless of the process:

- Project performance is collected and used to identify areas for improvement initiatives.
- There is a program in place to continuously collect and analyze process performance data and use it to improve the process.
- Lessons learned and best practices are used to improve the process.

2.3 PMBOK Knowledge Areas and Maturity Profile

The Project Management Institute (PMI) has published its standard for project management practice in a document entitled *A Guide to the Project Management Body of Knowledge* [3]. The current standard was published in 2000. PMBOK defines the project management life cycle in terms of five phases, or process groups, to use their terminology. They are initiating processes, planning processes, executing processes, controlling processes, and closing processes. Spread across these five process groups are 39 process areas grouped into nine knowledge areas (Figure 2.1), as described in the following sections. For each process there is a capsule description of the characteristics of that process at each maturity level. These descriptions will become the foundation of a survey instrument to assess the actual maturity level of the process and the maturity level of the practice of the process. The survey instrument is discussed in Chapter 3 and an example of an actual survey that was developed for a large retail client is given in the Appendix.

This section briefly describes each of the 39 project management processes grouped by knowledge area. Also, there is a more detailed characterization of each process at each maturity level. This information is the background used to develop the survey questions given in the Appendix.

2.3.1 Project Integration Management

Project integration management consists of three project management processes: project plan development, project plan execution, and integrated change control.

	Initiating	Planning	Executing	Controlling	Closing
Integration		Plan development	Plan execution	Integrated change control	
Scope	Initiation	Scope planning Scope definition		Scope verification Scope change control	
Time		Activity definition Activity sequencing Activity duration estimating Schedule development		Schedule control	
Cost		Resource planning Cost estimating Cost budgeting		Cost control	
Quality		Quality planning	Quality assurance	Quality control	
HR		Organizational planning Staff acquisition	Team development		
Comm.		Communications planning	Information distribution	Performance reporting	Administrative closure
Risk		Risk planning Risk identification Qualitative risk analysis Quantitative risk analysis Risk resource planning		Risk monitoring and control	
Procurement			Solicitation Source selection Contract administration		Contract Close-out

Figure 2.1 Processes cross-classified by process group and knowledge area.

2.3.1.1 Project Plan Development

Project plan development incorporates the output from other planning activities, which results in a coherent document that will guide the execution and control of the project. The process of accomplishing this is usually iterative. Iteration arises when the plan moves from generic resources and timelines to specific resources and timelines. Estimation, which is an integral part of the plan, will often evolve from range type estimates to more specific range or even point estimates of duration and cost. For a list of survey questions that applies to this process see pages 171–173 in the Appendix.

Project Plan Development—Level 1 Characteristics

- Each project manager has his/her own version of the project plan.
- A scope statement is prepared at the discretion of the project manager.
- A work breakdown structure (WBS) may be part of the project plan.
- There may be a work schedule with resource requirements.

- There may be milestones with scheduled dates.
- There will be few, if any, plan development standards.

Project Plan Development—Level 2 Characteristics

- There is a standard and documented process for developing the project plan.
- A WBS is part of the project plan.
- The project plan includes cost summary estimates.
- There is a work schedule with estimated costs and resource requirements.
- Milestones with scheduled dates are part of the project plan.
- Risk management and communications management are part of the project plan.
- A change control process is in place and used to update the project plan.

Project Plan Development—Level 3 Characteristics

- The project planning process is fully documented and implemented.
- All knowledge areas are planned and part of the project plan.
- Scope, time, cost, and risk are incorporated into the project plan at an appropriate level of detail.
- The project plan includes management of cost and schedule.
- A detailed WBS is developed to a level of detail that permits all cost and scheduling information to be developed and managed.

Project Plan Development—Level 4 Characteristics

- Project plans are integrated into other organizational strategic and operational processes and systems.
- The project plan integrates with corporate financial and related systems.
- There is a true integration of project plans and business plans.
- Lessons learned and best practices are captured and made available to other projects.

Project Plan Development—Level 5 Characteristics

- Processes are in place to continuously improve project plan development.
- There is a process in place to integrate project plans into organizational decision-making and decisions regarding the project portfolio.

- Lessons learned and best practices are used to improve the project plan development process.

2.3.1.2 Project Plan Execution

Project plan execution involves the project manager and the entire project team in the coordination and performance against plan of the work of the project. Monitoring and control tools will be in place to measure conformance against plan with corrective actions initiated to return the project to plan. For a list of survey questions that applies to this process see pages 173–174 in the Appendix.

Project Plan Execution—Level 1 Characteristics

- Work assignments are given informally and mostly through verbal communications.
- Verbal reports of work assignment results are common.

Project Plan Execution—Level 2 Characteristics

- Basic metrics tracking planned versus actual performance are reported.
- Cost and schedule information includes the impact of technical considerations in measuring project status.
- Summary information is compiled and integrated into milestone status reports.

Project Plan Execution—Level 3 Characteristics

- A project planning process is defined and documented and used by all projects.
- Project teams are providing project performance data on actual time, cost, and technical requirements.
- Performance data is integrated, analyzed, and reported.
- Deliverables status reports are provided.
- Cost and schedule variance reports are provided.
- Performance data from each of the knowledge areas is also aggregated and reported.

Project Plan Execution—Level 4 Characteristics

- Project status reports are integrated into other corporate level reports and processes.
- Corporate databases provide performance data on cost, time, and technical requirements, and reports are automatically produced.

- Knowledge area performance metrics data is integrated into other project performance reports.
- Lessons learned and best practices are captured and made available to other projects.

Project Plan Execution—Level 5 Characteristics

- Improvement processes are in place to capture best practices and lessons learned and feed them back into project management processes for their improvement.
- Project performance data is analyzed to understand if and how project management processes are being improved as evidenced by practice improvement.
- Lessons learned and best practices are used to improve the project plan development process.

2.3.1.3 Integrated Change Control

Change can occur either by choice or by circumstance. There must be a process in place to receive the change, assess the impact on the project plan and all related knowledge areas, act on the change, and revise all plans accordingly. For a list of survey questions that applies to this process see pages 174–175 in the Appendix.

Integrated Change Control—Level 1 Characteristics

- Changes are communicated to the project manager or to a team member in an informal manner.
- Teams use their own process for change control.

Integrated Change Control—Level 2 Characteristics

- There is a documented change control process in place.
- The change control process includes a change request form.
- Changes are monitored and logged.
- Project plans are updated as a result of approved change requests.

Integrated Change Control—Level 3 Characteristics

- All project teams utilize the defined and documented change control processes.
- Change control processes include scope, cost, and schedule changes.
- Baselines are established and managed.

Integrated Change Control—Level 4 Characteristics

- The change control processes are integrated into other organizational control processes.
- Data across all projects is monitored and controlled for all projects.
- Lessons learned and best practices are captured and made available to all projects.

Integrated Change Control—Level 5 Characteristics

- Continuous improvement processes are in place to analyze and act upon change control data.
- Change control data is analyzed to identify trends to improve project planning processes.
- Lessons learned and best practices are used to improve the integrated change control process.

2.3.2 Project Scope Management

Project scope management consists of five project management processes: initiation, scope planning, scope definition, scope verification, and scope change control.

2.3.2.1 Initiation

Initiation is a gating stage in the project. It is the formal authorization to proceed with the project or to take the project to the next phase. The decision is based on the demonstrated alignment of the project deliverables to the goals and objectives of the business unit for whom the project is being undertaken. The output from this process is a document called a project charter. It gives the project manager the formal authority to acquire and use organization resources for the work of the project. For a list of survey questions that applies to this process see page 175 in the Appendix.

Initiation—Level 1 Characteristics

- Projects are often informally initiated.
- The client does not have a formal statement of requirements but rather a general statement of intent.
- The technical requirements are also generally defined as a result of the client business requirements statement.
- There may be a list of deliverables.

Initiation—Level 2 Characteristics

- There is an approved statement of documented business requirements.
- A process of identifying and listing deliverables is in place.
- There is an approved statement of documented deliverables.

Initiation—Level 3 Characteristics

- A documented process is in place for identifying, documenting, and approving business requirements and is used by all projects.
- The process includes all relevant stakeholders and the project team in the formation and approval of business requirements.
- There is a documented process in place for defining technical requirements.
- The project team is involved in the definition, documentation, and approval of all technical requirements.
- Deliverables are defined and documented in detail with the cooperation of the client and the project team.

Initiation—Level 4 Characteristics

- Business requirements have been defined in the light of other related business processes.
- Technical requirements have been defined in the light of other related technical requirements.
- The deliverables list is constantly updated and maintained as other business and technical processes change.
- Lessons learned and best practices are captured and made available to other projects.

Initiation—Level 5 Characteristics

- Information on how business and technical processes have been defined and documented is used to improve the processes involved.
- Information on how the processes used to define and maintain the deliverables list is used to improve the processes involved.
- Lesson learned and best practices are used to continuously improve the initiation process.

2.3.2.2 Scope Planning

Scope planning follows directly from the initiation process. It involves the creation of a detailed scope statement agreed to by the customer and the project manager. This document becomes the foundation of the detailed project plan. For a list of survey questions that applies to this process see page 176 in the Appendix.

Scope Planning—Level 1 Characteristics

- There are no processes in place to guide in the development of a schedule or work plan.
- There may be a task list of the work to be done.

Scope Planning—Level 2 Characteristics

- There is a defined and documented process for creating the WBS.
- There is a process in place that defines how the project work schedule is developed directly from the WBS.
- Templates are defined and documented for scope planning.

Scope Planning—Level 3 Characteristics

- There is a scope management process defined and documented, which is used by all projects.
- A scope statement is produced by every project.

Scope Planning—Level 4 Characteristics

- The scope statement is considered in light of other business requirements.
- Lessons learned and best practices are captured and made available to other projects.

Scope Planning—Level 5 Characteristics

- A process is in place for the review of scope statements and their use in the continuous improvement of the scope planning process.

2.3.2.3 Scope Definition

Beginning with the scope statement, this process further defines the scope of the project by decomposing the scope and producing the WBS. The decomposition is also a validation that the project will deliver according to the client's request. This process is also a critical input to further processes that involve estimating cost, duration, and resource requirements needed to deliver the scope. For a list of survey questions that applies to this process see pages 176–177 in the Appendix.

Scope Definition—Level 1 Characteristics

- An ad hoc statement of project scope definition is prepared on a project by project basis.
- There is no consistent content or format for scope definition.

Scope Definition—Level 2 Characteristics

- There is a documented content and format for scope definition statements.
- Not all projects use the scope definition standards.
- The WBS is used as a format for scope definition.

Scope Definition—Level 3 Characteristics

- There is a documented and broadly approved scope definition in place that is used by all projects.
- An approved statement of work is in place.

Scope Definition—Level 4 Characteristics

- Scope statements are reviewed for their fit against business requirements.
- Interproject dependencies are considered in the review of scope statements.
- Lesson learned and best practices are captured and made available to other projects.

Scope Definition—Level 5 Characteristics

- There is a process in place for the continuous improvement of the scope definition process.
- Lessons learned and best practices are used to continuously improve the scope definition process.

2.3.2.4 Scope Verification

Scope verification is the formal acceptance by the sponsor and the client that the work results were as agreed to in the scope statement, the WBS, and the project plan. The successful completion of scope verification signals the beginning of the close-out process. For a list of survey questions that applies to this process see page 177 in the Appendix.

Scope Verification—Level 1 Characteristics

- There is no defined and documented process in place for scope verification.
- Scope verification is practiced at the discretion of the project manager.

Scope Verification—Level 2 Characteristics

- A process for using the WBS for scope verification has been defined and documented.
- There is a defined and documented formal process for scope verification acceptance.

Scope Verification—Level 3 Characteristics

- A formal acceptance of the scope statement is required for every project.

Scope Verification—Level 4 Characteristics

- The scope statement approval process includes some consideration of other business requirements.
- Lessons learned and best practices are captured and made available to other projects.

Scope Verification—Level 5 Characteristics

- There is a formal program in place for the continuous improvement of the scope verification process.
- Lessons learned and best practices are used for the continuous improvement of the scope verification process.

2.3.2.5 Scope Change Control

Scope change control includes the formal process of receiving change and change requests, evaluating the impact on the WBS and the project plan, and acting on the change. The output from this process will be an updated scope statement, the necessary corrective measures, lessons learned from the change, and the revised project schedule. For a list of survey questions that applies to this process see pages 178–179 in the Appendix.

Scope Change Control—Level 1 Characteristics

- There is no documented scope change control process.

- Changes are communicated in an ad hoc manner.
- Changes are irregularly documented at the preference of the project manager.

Scope Change Control—Level 2 Characteristics

- There is a documented scope change control process.
- Some projects use the documented scope control process.

Scope Change Control—Level 3 Characteristics

- All projects use the documented scope change control process.
- Project scope status is monitored and controlled.

Scope Change Control—Level 4 Characteristics

- All change control processes are integrated with other relevant organizational processes.
- Scope reporting is integrated into other project status reports.
- Lesson learned and best practices are captured and made available to other projects.

Scope Change Control—Level 5 Characteristics

- A process for the continuous improvement of the scope change control process is in place.
- Lessons learned and best practices are captured, documented, and distributed.
- Scope change performance is measured as an indication of project performance and effectiveness.

2.3.3 Project Time Management

Project time management consists of five project management processes: activity definition, activity sequencing, activity duration estimating, schedule development, and schedule control.

2.3.3.1 Activity Definition

Activity definition involves a further decomposition of the WBS from the deliverables format to the work that must be done to produce the deliverables identified in the WBS and the scope statement. For a list of survey questions that applies to this process see page 179 in the Appendix.

Activity Definition—Level 1 Characteristics

- Activity definition is ad hoc and may include a skeleton WBS with some milestone dates.
- There is no reason to believe that the milestone list is complete.
- Milestones may or may not be defined to the activity level.

Activity Definition—Level 2 Characteristics

- There is a documented process for activity and milestone definition.
- There is a documented process for scope statement preparation but project teams are not required to follow it.
- There is a standard process and template for the WBS.

Activity Definition—Level 3 Characteristics

- There is an organizational standard for activity definition, which all projects are required to use.
- There is an organizational standard for defining and scheduling detailed activities, which every project is required to use.
- The WBS is defined at least to the third level of detail.

Activity Definition—Level 4 Characteristics

- Management collects project performance data relevant to the activity schedule.
- Management decisions regarding a single project take into account the impact on related projects and other organizational processes.
- Lessons learned and best practices regarding activity definition and scheduling are captured and made available to other projects.

Activity Definition—Level 5 Characteristics

- A process is in place to continuously improve activity definition.
- Lesson learned and best practices are used to improve the activity definition process.

2.3.3.2 Activity Sequencing

The activity sequencing process is the beginning step to creating the project schedule. Laying out the sequence of what work can be done and when is the input needed to develop the project work schedule. For a list of survey questions that applies to this process see pages 180–181 in the Appendix.

Activity Sequencing—Level 1 Characteristics

- Project activities are sequenced on an ad hoc basis with little regard for dependencies.
- Software tools to support activity sequencing are seldom used.

Activity Sequencing—Level 2 Characteristics

- There is a documented process for activity sequencing and dependency determination.
- Network diagramming techniques exist.

Activity Sequencing—Level 3 Characteristics

- Activity sequencing processes using network templates are documented and standardized and used by all projects.
- Network diagrams are integrated into scheduling software.

Activity Sequencing—Level 4 Characteristics

- Interproject dependencies are monitored.
- Management decisions take into account the interdependencies between projects.
- Lessons learned and best practices are captured and made available for usage by other teams.

Activity Sequencing—Level 5 Characteristics

- There are documented processes in place to continuously monitor activity sequencing process performance and improve it.
- Lessons learned and best practices are captured and used to improve the activity sequencing processes.

2.3.3.3 Activity Duration Estimating

Activity duration estimating produces a duration estimate for every work activity that will produce all or some part of a deliverable. These estimates can be generated in any one of several ways that include expert judgment, analogous estimating, quantitative models, and others. These estimates may evolve through several stages as more information is discovered during project work. For a list of survey questions that applies to this process see page 182 in the Appendix.

Activity Duration Estimating—Level 1 Characteristics

Activity duration estimating is an informal process.

Activity Duration Estimating—Level 2 Characteristics

- There is a defined and documented process for activity duration estimating.
- The process for estimating activity duration is based on historical information, industry standards, and expert knowledge.
- Estimated and actual activity durations are archived for use in other projects.

Activity Duration Estimating—Level 3 Characteristics

- There is a defined and documented process in place for activity duration estimating which is used by all projects.
- Alternative estimating procedures are documented and available for team usage.
- Actual activity duration data is collected, analyzed, and used for similar activity duration estimates.

Activity Duration Estimating—Level 4 Characteristics

- Historical data on estimated and actual activity duration is available for team usage.
- Lessons learned and best practices are captured and made available to other projects.

Activity Duration Estimating—Level 5 Characteristics

- There is a program in place for the continuous improvement of the activity duration estimating process.
- Lessons learned and best practices are used to continuously improve the activity duration estimating process.

2.3.3.4 Schedule Development

The schedule development process develops the estimated start and end dates for every work activity needed to produce the deliverables identified in the WBS. The primary input data needed to create this timed schedule is the project network diagram and activity duration estimates. For a list of survey questions that applies to this process see pages 182–184 in the Appendix.

Schedule Development—Level 1 Characteristics

- There is no network-based schedule development process in place.
- Schedule development is done on an ad hoc basis by project teams.

- High-level estimates (i.e., milestone dates) are crudely estimated or rough guesses at best.

Schedule Development—Level 2 Characteristics

- A complete documented process for schedule development exists.
- A process for managing the schedule plan exists.
- Staffing plans are developed and coordinated with management.
- Project schedules are developed based on resource availability.
- A variety of scheduling methods are available to the teams.

Schedule Development—Level 3 Characteristics

- There is a standard set of project management software tools that all projects are using.
- There is a documented standard for schedule development processes that all projects are using.

Schedule Development—Level 4 Characteristics

- Management uses schedule performance data as an input to decision making activities.
- A process is documented and in place to inform management with respect to schedule status.
- Lessons learned and best practices are collected and utilized.

Schedule Development—Level 5 Characteristics

- A process is documented and in place for the continuous improvement of schedule development.
- Lessons learned and best practices are captured and utilized for improvement of the schedule development process.

2.3.3.5 Schedule Control

The schedule will change several times throughout the life cycle of the project. Some of these changes are the result of events exogenous to the project. Others are client induced. In both cases the changes will impact the schedule and must be managed. Schedule changes must be integrated into several other processes as pointed out in the project integration management process. For a list of survey questions that applies to this process see pages 184–186 in the Appendix.

Schedule Control—Level 1 Characteristics

- Schedules are managed and controlled at the project level using whatever means they select.

- Schedule performance tracking is inconsistent.
- Schedule performance reports are prepared in an ad hoc manner and are not consistent across projects.
- There is no documented approach to schedule control.

Schedule Control—Level 2 Characteristics

- A process is documented for the control of schedules including a change control process.
- A schedule reporting system exists.
- Schedule baselines exist, as do planned versus actual reports.

Schedule Control—Level 3 Characteristics

- There is a documented standard for change control, schedule reporting, and cost/schedule control that is used on all projects.
- Schedule performance against plan is monitored and managed.

Schedule Control—Level 4 Characteristics

- A comprehensive cost/schedule control system is used for management decisions.
- Lessons learned and best practices are captured and made available to other teams.

Schedule Control—Level 5 Characteristics

- A process is in place to capture schedule performance and improve the process.
- Lessons learned and best practices are used for schedule control process improvement.

2.3.4 Project Cost Management

Project cost management consists of four project management processes: resource planning, cost estimating, cost budgeting, and cost control.

2.3.4.1 Resource Planning

The resources required to complete the work of the project are the focus of the resource planning process. The resources include people, materials, and equipment. The planning activities include estimating what resources are needed, in what quantities, and when. Oftentimes resource availability may compromise the schedule and schedule changes will be required. For a list of survey questions that applies to this process see pages 186–188 in the Appendix.

Resource Planning—Level 1 Characteristics

- Each project manager employs their own approach to identifying resource types and quantities.
- The ad hoc processes employed do not always result in a complete accounting of needed resources.

Resource Planning—Level 2 Characteristics

- Listings of all resources (labor, equipment, facilities) are available.
- There is a process in place to conduct resource planning.
- There is a process and templates are in place for planning resource requirements.
- There is a process and templates are in place to estimate resource requirements.

Resource Planning—Level 3 Characteristics

- All project teams are utilizing a defined and documented process for resource planning.
- A process is in place for managing resource requirements vis-à-vis resource availability.

Resource Planning—Level 4 Characteristics

- Resource planning processes are integrated into other business processes and practices.
- Lessons learned and best practices are captured and made available to other projects.

Resource Planning—Level 5 Characteristics

- There is a program in place for the continuous improvement of the resource planning process.
- Lessons learned and best practices are used for the continuous improvement of the resource planning process.

2.3.4.2 Cost Estimating

The cost estimating process is driven by the resource planning process. Using the same tools that were used to estimate duration, the planning team will generate estimates of cost. For many of the work activities these costs will be calculated using standard costing tables rather than creating original estimates of cost. For a list of survey questions that applies to this process see pages 188–189 in the Appendix.

Cost Estimating—Level 1 Characteristics

- Cost estimates are done in an ad hoc manner and do not necessarily contain all relevant areas where costs may be accrued.
- There is no organizational guideline for preparing and documenting cost estimates.
- Some tools and templates exist but they are by no means complete or fully documented for use.

Cost Estimating—Level 2 Characteristics

- There is a documented process for preparing and documenting cost estimates.
- A documented process exists that contains templates that relate the WBS to summary level estimates of cost.
- A cost management plan and process for managing cost through the project exists.
- There are a number of documented tools, cost standards, commercial databases, templates, and estimating processes.
- A process exists for collecting and reporting actual costs.

Cost Estimating—Level 3 Characteristics

- A complete cost estimating and analysis system is in place, is documented, and is being used on all projects.
- Cost standards are documented and being used on all projects.
- Analyses of budgeted versus actual are done routinely on all projects.
- A historical database of estimated cost versus actual cost is maintained for use in future projects.

Cost Estimating—Level 4 Characteristics

- Cost estimating is fully integrated into other business processes.
- Organizational cost standards are in place and consistently used.
- Lessons learned and best practices are captured and made available to other projects.

Cost Estimating—Level 5 Characteristics

- A program for the continuous improvement of the cost estimating process is documented and being used.
- Lessons learned and best practices are used to improve cost estimating procedures.

2.3.4.3 Cost Budgeting

Once cost estimation is complete and approved, monies can be allocated to the various work activities. These allocations will be used later to measure project performance. For a list of survey questions that applies to this process see pages 189–190 in the Appendix.

Cost Budgeting—Level 1 Characteristics

- There is no established process for budgeting costs.
- Some project teams have developed their own approach to cost budgeting that are specific to their project and not reflective of any existing processes.

Cost Budgeting—Level 2 Characteristics

- There is a documented process for preparing the project budget.
- Baseline budgets are in place in most projects.

Cost Budgeting—Level 3 Characteristics

- Project budget baselines are established and managed in accordance with standardized budgeting processes.
- The budgeting process is fully integrated with the project scheduling system.

Cost Budgeting—Level 4 Characteristics

- The cost budgeting process is fully documented and in use by all projects.
- Lessons learned and best practices are captured and made available to other projects.
- The cost budgeting process is fully integrated in other organizational processes and systems.

Cost Budgeting—Level 5 Characteristics

- A process to continuously improve the cost budgeting process is in place and being used.
- Lessons learned and best practices are used to improve the cost budgeting process.

2.3.4.4 Cost Control

Because the schedule will change several times throughout the life cycle of the project, costs can also be expected to change. These changes will impact the project budget as well as the allocations to the various work activities. These cost changes must be managed. Cost changes must be integrated into several other processes as pointed out in the project integration management process. For a list of survey questions that applies to this process see pages 190–192 in the Appendix.

Cost Control—Level 1 Characteristics

- The control and management of cost is left up to the discretion of the individual project managers.
- Cost reports are provided on an ad hoc basis and do not follow any standard template.

Cost Control—Level 2 Characteristics

- There is a standard format in place for reporting cost reports.
- A process is in place for managing and controlling cost.
- There is a cost change management process.
- Estimated and actual costs are reported by project teams.

Cost Control—Level 3 Characteristics

- A fully documented cost control process is in place and is being used by all project teams.
- Cost and schedule reports are integrated.
- Baselines, reports, earned value, schedule variances, and performance metrics are in place and being used to monitor and report project progress and status.

Cost Control—Level 4 Characteristics

- The cost control process is integrated with the organization's other control processes.

- Actual costs are compared with budgeted costs to evaluate project cost performance and status.
- Lessons learned and best practices are captured and made available to other teams.

Cost Control—Level 5 Characteristics

- A process for the continuous improvement of the cost control is in place and being used.
- Lessons learned and best practices are being used to improve the cost control process.

2.3.5 Project Quality Management

Project quality management consists of three project management processes: quality planning, quality assurance, and quality control.

2.3.5.1 Quality Planning

Operating from a stated quality policy, the project leadership must determine which quality standards are appropriate for this project and how they will meet those standards. For a list of survey questions that applies to this process see pages 192–193 in the Appendix.

Quality Planning—Level 1 Characteristics

- Quality planning is done on an ad hoc basis at the discretion of the project manager.
- There is no defined and documented quality planning process in place.

Quality Planning—Level 2 Characteristics

- A defined and documented quality planning process exists.
- Metrics exist for comparing project quality performance against organizational quality standards.

Quality Planning—Level 3 Characteristics

- A comprehensive quality planning process is in place, documented, and is being used by all projects.
- The organization has defined positions that are responsible for organizational quality standards and assurance.

Quality Planning—Level 4 Characteristics

- Quality planning is an integrated enterprise-wide process.

- There is an independent quality office.
- Lessons learned and best practices are captured and available to other teams.

Quality Planning—Level 5 Characteristics

- A process is in place and being used to improve quality planning.
- Lessons learned and best practices are being captured and used to improve the quality planning process.

2.3.5.2 Quality Assurance

Quality assurance consists of the processes and procedures activated within the quality plan to assure that the relevant quality standards are in fact met. These processes and procedures should be under the administration and control of an outside unit to assure an unbiased verification. For a list of survey questions that applies to this process see pages 194–195 in the Appendix.

Quality Assurance—Level 1 Characteristics

- Quality assurance processes are defined at the individual project level at the discretion of the project manager.
- There are no established quality assurance processes in place.

Quality Assurance—Level 2 Characteristics

- There is a defined and documented process for quality assurance.
- A number of quality assurance tools and techniques are used by some project teams but not as part of any organized and standardized quality effort.
- Project teams have developed quality checklists are part of the project review process.

Quality Assurance—Level 3 Characteristics

- A number of quality tools and techniques are defined and documented and considered part of a standardized approach used by all projects.
- Project reviews and walkthroughs include a review of quality efforts.

Quality Assurance—Level 4 Characteristics

- Project reviews are conducted to assure that the deliverables meet business requirements.

- Metrics are defined and collected and compared to industry standards to identify quality performance problems.
- Lessons learned and best practices are captured and made available to other teams.

Quality Assurance—Level 5 Characteristics

- Processes are in place to continuously improve quality management processes.
- Lessons learned and best practices are used to improve project quality management processes.

2.3.5.3 Quality Control

The quality control process is the administrative part of quality management. The project deliverables and project management process is monitored to determine compliance and to correct any anomalies. For a list of survey questions that applies to this process see pages 195–196 in the Appendix.

Quality Control—Level 1 Characteristics

- Practices and standards for quality control do not exist.
- Some teams will put ad hoc procedures in place to monitor the quality of their work.
- Product testing is done informally by the developers.

Quality Control—Level 2 Characteristics

- Quality control processes are defined and documented.
- Quality control tools include acceptance criteria, performance standards, business requirements, and quality standards for review and testing.
- Various testing procedures are defined such as subsystem testing, unit testing, integration testing, and final product testing.
- Templates and guidelines exist and are used in the quality process.
- Project review metrics are defined for quality acceptance criteria and performance requirements.

Quality Control—Level 3 Characteristics

- Standardized quality control processes and tools are defined and documented and being used by all projects.

- Clients are meaningfully involved in test scenario definition and execution.
- Product testing is done in a simulated environment.
- Projects are using templates and guidelines for integration testing.
- Projects are using templates and guidelines for acceptance testing.
- The client is actively involved in integration testing.
- Acceptance testing and approval is driven by the client.

Quality Control—Level 4 Characteristics

- Project deliverables are tested against organizational quality standards.
- Processes are in place for integration testing of those deliverables that interact with other organizational products and processes.
- Lessons learned and best practices are captured and made available to other project teams.

Quality Control—Level 5 Characteristics

- Processes are in place for the continuous improvement of quality control processes.
- Lesson learned and best practices are used to improve the quality control process.

2.3.6 Project Human Resources Management

Project human resources management consists of three project management processes: organizational planning, staff acquisition, and team development.

2.3.6.1 Organizational Planning

The organizational planning process focuses on establishing and documenting the project infrastructure of positions, roles, reporting relationships, and responsibilities for project teams. For a list of survey questions that applies to this process see pages 197–198 in the Appendix.

Organizational Planning—Level 1 Characteristics

- There are no standardized processes for determining staffing requirements.
- Project managers are assigned on an ad hoc basis.
- Project managers use ad hoc procedures to determine the staffing requirements for their projects.

- Project team members receive their assignments directly from the project manager.

Organizational Planning—Level 2 Characteristics

- There is a documented process for determining staffing requirements.
- Project managers document their skill requirements and project organizational chart based on their own preferences.
- The staffing plan accounts for corrective steps as the project progresses.
- Project team position responsibilities are provided by the project manager.

Organizational Planning—Level 3 Characteristics

- A process is defined and documented for analyzing project staffing requirements in light of organizational constraints and other project staffing requirements and is used by all projects.
- Position descriptions exist for all project personnel.

Organizational Planning—Level 4 Characteristics

- Project staffing processes are integrated into organizational staffing processes.
- Staffing decision processes take into account the impact on the organization of various staffing alternatives.
- Lessons learned and best practices are captured and made available to other teams.

Organizational Planning—Level 5 Characteristics

- Processes are in place to continuously improve organizational planning processes.
- Metrics have been established and are being used to collect data to measure the effectiveness of organizational planning processes.
- Lessons learned and best practices are used to improve organizational planning processes.

2.3.6.2 Staff Acquisition

The staff acquisition process responds to the project plan to provide the requisite human resources to meet the project work schedule. This involves finding the most skilled combination of people based on their availabilities to staff projects that are commissioned for action. For a list of survey questions that applies to this process see pages 198–199 in the Appendix.

Staff Acquisition—Level 1 Characteristics

- There are no documented processes in place for acquiring staff.
- Project managers deal on an ad hoc basis with other managers to recruit their team members.

Staff Acquisition—Level 2 Characteristics

- There is an ad hoc process being followed to determine who is available to work on a specific project.
- Acquiring a specific person for a project team is a matter of negotiations between the project manager and the person's manager.
- A process is defined and documented for acquiring project staff.
- Project managers request project staff be allocated by their managers for certain time periods.

Staff Acquisition—Level 3 Characteristics

- There is a process in place and used by all projects to request project staffing from an organizational pool under management control.
- Negotiations for specific resources are commonly done between project managers and resource pool managers.
- When necessary, the project manager looks outside the organization for staffing resources.

Staff Acquisition—Level 4 Characteristics

- There is a process in place for prioritizing staffing requirements and assigning staff to projects.
- Lessons learned and best practices are captured and made available to other teams.

Staff Acquisition—Level 5 Characteristics

- A process is in place and being used to continuously improve staff acquisition processes.
- Resource forecasting is used at the corporate level.
- Lessons learned and best practices are being used to continuously improve staff acquisition processes.

2.3.6.3 Team Development

The team development process focuses on developing the skills and competencies of all team members from stakeholders to individual team members. The goal is to configure a team whose likelihood of success is optimal given the resource constraints and the specific needs of the project team. For a list of survey questions that applies to this process see pages 200–202 in the Appendix.

Team Development—Level 1 Characteristics

- There are no processes defined and documented for conducting team development programs.
- The project manager employs ad hoc processes and practices to create a sense of team among the project staff.
- Team meetings are often held.

Team Development—Level 2 Characteristics

- A process is in place to get a newly formed team started on a new project.
- Processes are in place to incorporate the team in planning activities.
- Project reviews, status meetings, walkthroughs, and a variety of other team-related activities are in place and used on an as-needed basis.
- A performance review process is in place for project managers to review team member performance and recommend rewards and recognition.
- Team processes are in place for conflict management, problem resolution, and decision-making.

Team Development—Level 3 Characteristics

- Team processes are documented and being used by all projects.
- Management is fully involved in team development activities.

Team Development—Level 4 Characteristics

- Organization-wide team development processes are in place and fully integrated with projects.
- Training and development needs are determined and communicated to the proper management.
- Performance evaluations include evaluations by the project manager.
- Lessons learned and best practices are captured and made available to project teams.

Team Development—Level 5 Characteristics

- Feedback from teams who have completed projects is used to improve the process of team development.
- Lessons learned and best practices are used to improve team development processes.

2.3.7 Project Communications Management

Project communications management consists of four project management processes: communications planning, information distribution, performance reporting, and administrative closure.

2.3.7.1 Communications Planning

Communications planning focuses on defining who the stakeholders are who will require information on the project, determining what their information needs are, how often they will need that information, and in what format the information should be presented to them. This initial needs analysis is done as part of project planning and continues through the life cycle of the project. For a list of survey questions that applies to this process see pages 202–203 in the Appendix.

Communications Planning—Level 1 Characteristics

- There are no communications planning processes documented or defined.
- Project teams develop their own ad hoc communications plans.
- The project manager provides project status reports only when requested.

Communications Planning—Level 2 Characteristics

- A process for stakeholder analysis is defined and documented.
- A process for generating communications plans from the stakeholder analysis is defined.
- A variety of status reports are developed and available for use.

Communications Planning—Level 3 Characteristics

- All projects are generating a communications plan following a documented standard process.
- The communications plan accounts for varying needs of stakeholders.

Communications Planning—Level 4 Characteristics

- The communications plan is integrated into other corporate systems.
- Lessons learned and best practices are captured and made available to other projects.

Communications Planning—Level 5 Characteristics

- Feedback and analysis of communications plans are used to continuously improve the communications planning process.
- Lessons learned and best practices are used to improve the communications planning process.

2.3.7.2 Information Distribution

This process implements the communications plan and also prepares for any unexpected requests for project information. For a list of survey questions that applies to this process see pages 203–204 in the Appendix.

Information Distribution—Level 1 Characteristics

- There is no process in place that defines the information distribution.
- Project teams develop their own ad hoc procedure for information distribution.
- Project teams respond to requests for information in an ad hoc manner.

Information Distribution—Level 2 Characteristics

- Information distribution processes and media are defined and documented.
- A project information system is operational for the storage and retrieval of project performance data and other relevant information.

Information Distribution—Level 3 Characteristics

- Processes are defined in which stakeholders can access and retrieve information on projects of interest.
- All teams are using a standardized system for collecting and distributing project information.

Information Distribution—Level 4 Characteristics

- Information collection and distribution is integrated with other corporate information systems.

- Lessons learned and best practices are captured and made available to other projects.

Information Distribution—Level 5 Characteristics

- Processes are in place for analyzing information distribution and distribution systems for improving the information distribution process.
- Lessons learned and best practices are used to improve the information distribution process.

2.3.7.3 Performance Reporting

This process collects and disseminates project performance data. It includes analyses of how resources, budget, and time are being used to achieve project objectives. Status reports, progress reports, and forecasts are typical deliverables from this process. For a list of survey questions that applies to this process see pages 204–205 in the Appendix.

Performance Reporting—Level 1 Characteristics

- There is no process in place for performance reporting.
- Project managers respond in an ad hoc manner to requests for project performance information.

Performance Reporting—Level 2 Characteristics

- A comprehensive summary reporting process is defined for cost, schedule, and performance and is available to projects.
- Estimated versus actual cost and time data can be collected according to a defined process.
- Milestone attainment is reported.
- An acceptance criteria procedure is in place and available for projects.

Performance Reporting—Level 3 Characteristics

- Processes are in place and required for reporting cost and schedule performance in a standardized manner so projects can be compared.
- Performance reviews are conducted.
- Project reports are archived for future use.

Performance Reporting—Level 4 Characteristics

- Performance data is collected and integrated into other corporate systems.

- Lessons learned and best practices are captured and made available to other projects.

Performance Reporting—Level 5 Characteristics

- Processes are in place and used to capture and analyze performance reporting data to improve the performance reporting process.
- Performance metrics are used to define efficiency and effectiveness as indicators of potential improvement opportunities.
- Lessons learned and best practices are used to improve the performance reporting process.

2.3.7.4 Administrative Closure

The project deliverables have been provided or the project has been terminated prior to completion and certain closing activities must now be completed. These activities will include final reporting, accumulation and distribution of lessons learned, documentation of final project results, and archiving of this information. For a list of survey questions that applies to this process see page 205 in the Appendix.

Administrative Closure—Level 1 Characteristics

- There are no defined processes for administrative closure.
- Project teams develop ad hoc administrative closure activities on a project-by-project basis.
- Final reports are completed at the discretion of the project manager.

Administrative Closure—Level 2 Characteristics

- There is a defined and documented process for project closure.
- Templates are available for the final project report.
- There is a process for collecting and archiving lessons learned.

Administrative Closure—Level 3 Characteristics

- All projects follow a defined and documented process for project closure.
- Final project reports are archived for future use.

Administrative Closure—Level 4 Characteristics

- There is a formal process for validating that project success criteria have been met.

- Lessons learned and best practices are captured and made available to all project teams.

Administrative Closure—Level 5 Characteristics

- There is a program in place to review administrative closure for possible improvement opportunities.
- Lessons learned and best practices are used to improve the administrative closure process.

2.3.8 Project Risk Management

Project risk management consists of six project management processes: risk management planning, risk identification, qualitative risk analysis, quantitative risk analysis, risk response planning, and risk monitoring and control.

2.3.8.1 Risk Management Planning

This process begins the risk management program by planning exactly how risk management will be approached. For a list of survey questions that applies to this process see pages 205–206 in the Appendix.

Risk Management Planning—Level 1 Characteristics

- There is no defined and documented process in place for risk management planning.
- Risk management planning is left to the discretion of the project manager.

Risk Management Planning—Level 2 Characteristics

- There is a defined and documented process for risk management planning.
- There is a documented risk management policy at the organizational level.

Risk Management Planning—Level 3 Characteristics

- All projects are using the standard risk management planning process.
- There is a formal risk analysis process that is required of all projects.

Risk Management Planning—Level 4 Characteristics

- There is a formal risk review process used by all projects.

- Risk analysis is used to prioritize all projects in the organization's portfolio.
- Lessons learned and best practices are captured and made available to other projects.

Risk Management Planning—Level 5 Characteristics

- There is a program in place for the continuous improvement of the risk management planning process.
- Lessons learned and best practices are used for the improvement of the risk management planning process.

2.3.8.2 Risk Identification

Risk identification is an iterative process that involves first the project team, then stakeholders and other managers affected by or affecting the project, and finally outside individuals who can comment on completeness based on their similar experiences. For a list of survey questions that applies to this process see pages 206–208 in the Appendix.

Risk Identification—Level 1 Characteristics

- There is no defined and documented process for identifying risks.
- Project team members occasionally suggest potential risks to the project manager.
- Project teams initiate risk discussions on an as-needed and when-needed basis.

Risk Identification—Level 2 Characteristics

- There is a defined and documented process for risk identification.
- The project team will examine the WBS, cost, schedule, and other relevant aspects of the project plan to identify the operative risks.
- Risk identification includes input from clients and stakeholders.
- Risk discussion includes cost, schedule, and scope.
- The project team may rely on industry lessons to identify risks.

Risk Identification—Level 3 Characteristics

- There is a documented standardized risk identification process in place that is used by all projects.
- There is a historical database of risks that project teams can use as a template.

- Interproject risks are identified.

Risk Identification—Level 4 Characteristics

- The risk identification process is fully integrated into other corporate processes and procedures.
- Lessons learned and best practices are captured and made available to other projects.

Risk Identification—Level 5 Characteristics

- A program is in place for the continuous collection and analysis of risk identification process performance data and used to improve the process.
- Lessons learned and best practices are used to improve the risk identification process.

2.3.8.3 Qualitative Risk Analysis

The qualitative risk analysis process prioritizes the identified risks and assesses their impact on the project and the likelihood that the risks will materialize. This process should be repeated through the life cycle of the project. It can inform the project plan as to suggested revisions. For a list of survey questions that applies to this process see page 208 in the Appendix.

Qualitative Risk Analysis—Level 1 Characteristics

- There is no defined and documented process for qualitative risk analysis.
- Qualitative risk analysis is done at the discretion of the project manager.
- Ad hoc qualitative risk analysis is done on a project-by-project basis and does not conform to any standard approach.

Qualitative Risk Analysis—Level 2 Characteristics

- There is a defined and documented process for qualitative risk analysis.
- Templates and tools are available for qualitative risk analysis.

Qualitative Risk Analysis—Level 3 Characteristics

- All projects are using a standardized approach to qualitative risk analysis.
- All projects are required to use a standardized risk matrix.

Qualitative Risk Analysis—Level 4 Characteristics

- There is an integrated process for qualitative risk analysis at the corporate level.
- There is a corporate level process for ranking projects based on risk.
- Lessons learned and best practices are captured and made available to other projects.

Qualitative Risk Analysis—Level 5 Characteristics

- There is a program in place for the continuous improvement of the qualitative risk analysis process.
- Lessons learned and best practices are used for the continuous improvement of the qualitative risk analysis process.

2.3.8.4 Quantitative Risk Analysis

This process analyses the probability of risk occurrence and the cost to the project. The result is an assessment of the priorities of each risk as related to their impact on project success. For a list of survey questions that applies to this process see pages 208–209 in the Appendix.

Quantitative Risk Analysis—Level 1 Characteristics

- There is no standardized approach to quantitative risk analysis.
- The project manager speculates on the potential impact of a risk.
- Risk discussions do not include any quantitative component.

Quantitative Risk Analysis—Level 2 Characteristics

- There is a standardized approach to quantitative risk analysis.
- Risk analysis includes a monetary impact along with the probability of the occurrence of the risk.
- Risks are evaluated on a project-by-project basis using a common methodology.
- Risks are prioritized on a single criterion.

Quantitative Risk Analysis—Level 3 Characteristics

- All projects use a standard approach to quantitative risk analysis.
- Multiple criteria are used to prioritize risk.

Quantitative Risk Analysis—Level 4 Characteristics

- Risk quantification takes into account organizational considerations.

- Risks are evaluated on an organizational basis.
- Performance indexes are taken into account when calculating risk impact.
- Lessons learned and best practices are captured and made available to other projects.

Quantitative Risk Analysis—Level 5 Characteristics

- There is a program in place for the continuous improvement of the quantitative risk analysis process.
- Risks are input to management for project related decisions.
- Lessons learned and best practices are used to improve the quantitative risk analysis process.

2.3.8.5 Risk Response Planning

This planning process focuses on reducing the likelihood of adverse effects on the project by identifying appropriate actions to be taken for each risk for which a response is advised. The determination of whether or not to respond is based on an analysis of the cost of response versus the loss to be expected from the risk. If the expected loss exceeds the cost of the response, then a risk response will be documented for action. For a list of survey questions that applies to this process see pages 209–211 in the Appendix.

Risk Response Planning—Level 1 Characteristics

- Risks are considered as they occur.
- There is no process in place for risk response planning.

Risk Response Planning—Level 2 Characteristics

- There is a defined and documented risk response planning process.
- Project team members discuss their strategy for dealing with risks.
- The organization requests teams to develop a plan to manage risks.

Risk Response Planning—Level 3 Characteristics

- All projects are using a standardized risk response planning process.
- The risk response planning process includes planning templates.
- Contingency plans and mitigation strategies are in place for every risk item.

Risk Response Planning—Level 4 Characteristics

- The risk response planning process is integrated into other corporate planning processes.
- Risk response planning is integrated into the cost and schedule planning processes.
- Lessons learned and best practices are captured and made available to other projects.

Risk Response Planning—Level 5 Characteristics

- There is a program in place to continuously improve the risk response planning process.
- Project reserves are available and managed with respect to their use for risk mitigation.
- Lessons learned and best practices are used to improve the risk response planning process.

2.3.8.6 Risk Monitoring and Control

This process maintains the risk plan through the project life cycle. Risks change, new risks appear, and other risks become inoperative as the project commences. The process also includes a monitoring function to assess risk reduction effectiveness. For a list of survey questions that applies to this process see pages 211–212 in the Appendix.

Risk Monitoring and Control—Level 1 Characteristics

- There is no defined and documented process in place for risk monitoring and control.
- Project teams create ad hoc processes for observing risk-related events and solving any risk-related problems as they occur.

Risk Monitoring and Control—Level 2 Characteristics

- There is a defined and documented process for risk monitoring and control.
- Responsibility for each risk item is assigned as they occur.
- Risk items are discussed in team meetings.
- A risk log is created and maintained by the team.
- Risk status is reported to stakeholders.

Risk Monitoring and Control—Level 3 Characteristics

- All teams are using a standardized risk monitoring and control process.
- Project risks are tracked and reported on an ongoing basis.

Risk Monitoring and Control—Level 4 Characteristics

- Risk monitoring and control is integrated into other organizational processes.
- Lessons learned and best practices are captured and made available to other projects.

Risk Monitoring and Control—Level 5 Characteristics

- There is a program in place for the continuous improvement of the risk monitoring and control process.
- Risk assessments are included in the determination of process effectiveness and efficiency.
- Lesson learned are used to continuously improve the risk monitoring and control process.

2.3.9 Project Procurement Management

Project procurement management consists of six project management processes: procurement planning, solicitation planning, solicitation, source selection, contract administration, and contract close-out.

2.3.9.1 Procurement Planning

Procurement planning involves identifying which project products or services can be provided by an outside agent. The plan would identify these needs and define how, what, and when to procure. For a list of survey questions that applies to this process see pages 212–214 in the Appendix.

Procurement Planning—Level 1 Characteristics

- There is no defined or documented process for procurement planning.
- Project managers will put together an approach for acquiring services and products from outside vendors on an as-needed basis.

Procurement Planning—Level 2 Characteristics

- A procurement planning process is defined and documented and available for projects.

- A statement of work is prepared by project teams for acquiring outside products and services.
- A procurement plan is prepared by the project manager that details what is to be procured, what should be paid, the specifications of what is needed, and a schedule for acquisition.

Procurement Planning—Level 3 Characteristics

- There is a process that teams are using to submit procurement recommendations to management.
- The decision to acquire outside products and services takes into account organizational considerations as well as project considerations.

Procurement Planning—Level 4 Characteristics

- Acquisition decisions are made at the organizational level.
- Lessons learned and best practices are captured and made available to projects.

Procurement Planning—Level 5 Characteristics

- There is a program in place that continuously collects and analyzes procurement planning performance data in order to improve the procurement planning process.
- Lessons learned and best practices are used to improve the procurement planning process.

2.3.9.2 Solicitation Planning

Once the decision has been made to acquire some product or service from outside the organization, the next step is to plan for that activity. The solicitation plan will include preparation of the appropriate forms and documents, evaluation criteria, and the statement of work with updates. For a list of survey questions that applies to this process see pages 214–216 in the Appendix.

Solicitation Planning—Level 1 Characteristics

- There is no defined and documented process for solicitation planning.
- Solicitation planning is done at the discretion of the project manager using ad hoc procedures.

Solicitation Planning—Level 2 Characteristics

- There is a defined process for solicitation planning.

- There is a procurement organization that takes the lead in identifying potential vendors.
- There is a process in place for identifying contract requirements.
- A clear evaluation criterion is developed for prioritizing vendor responses.
- The project team is meaningfully involved in establishing procurement documentation.
- RFPs are prepared for nonstandard and high cost products and services.

Solicitation Planning—Level 3 Characteristics

- All teams are using a standardized process for solicitation planning.
- There is a preferred bidder list.
- There are procurement templates to assist in documentation creation.
- A variety of contract types exist for procuring outside products or services.

Solicitation Planning—Level 4 Characteristics

- Project requisitions are fully integrated in the organization's requisition process.
- Lessons learned and best practices are captured and made available to other projects.

Solicitation Planning—Level 5 Characteristics

- There is a program in place to continuously improve the solicitation planning process.
- Solicitation planning is continuously monitored for efficiency and effectiveness.
- Lessons learned and best practices are used to improve the solicitation planning process.

2.3.9.3 Solicitation

This process is intended to elicit proposals and bids from prospective vendors. For a list of survey questions that applies to this process see pages 216–217 in the Appendix.

Solicitation—Level 1 Characteristics

- There is no defined and documented process for solicitation.

- Project teams follow an informal process for contacting potential vendors.

Solicitation—Level 2 Characteristics

- There is a defined and documented process in place for gathering information from the industry.
- The procuring unit contacts vendors for price comparisons and delivery dates.
- Vendors are required to submit a plan for delivery of products and services.
- The project manager collaborates with the procuring unit regarding vendor evaluation.

Solicitation—Level 3 Characteristics

- All projects are following a standard solicitation process.
- Vendors must comply with established project management processes.
- Vendors must submit a detailed project plan.
- Solicitation is done with the involvement of the project team, the procurement unit, and the legal department.

Solicitation—Level 4 Characteristics

- The solicitation process is fully integrated into other organizational processes and procedures.
- Lessons learned and best practices are captured and made available to other projects.

Solicitation—Level 5 Characteristics

- There is a program in place to continuously improve the solicitation process.
- Vendors are evaluated for efficiency and effectiveness.
- Lessons learned and best practices are used to improve the solicitation process.

2.3.9.4 Source Selection

This process focuses on the receipt of responses and the selection of the vendor based on the application of the evaluation criteria developed earlier. For a list of survey questions that applies to this process see page 217 in the Appendix.

Source Selection—Level 1 Characteristics

- There is no defined or documented process for source selection.
- The project manager follows ad hoc processes for source selection.

Source Selection—Level 2 Characteristics

- There is a defined and documented process for source selection.
- The project manager may have a personal list of vendors he is familiar with and may select from that list.

Source Selection—Level 3 Characteristics

- All projects are using a standard approach for source selection.
- There is an approved list of vendors that is used by projects.

Source Selection—Level 4 Characteristics

- The organization can leverage numerous requests to the same vendor.
- Lessons learned and best practices are captured and made available to other projects.

Source Selection—Level 5 Characteristics

- There is a program to continuously improve the source selection process.
- Lessons learned and best practices are used to improve the source selection process.

2.3.9.5 Contract Administration

A legal contract will have been negotiated with the vendor(s). This process monitors vendor performance relative to the contract provisions and assures compliance. For a list of survey questions that applies to this process see pages 217–219 in the Appendix.

Contract Administration—Level 1 Characteristics

- There is no defined and documented process in place for contract administration.
- Project managers deal directly with vendors with a contract that has few requirements beyond delivery dates.

Contract Administration—Level 2 Characteristics

- There is a defined and documented process available for contract administration.

- Change management processes are used to manage changes to contracts.
- Status and performance reports are prepared and distributed.
- Acceptance criteria are in place.

Contract Administration—Level 3 Characteristics

- All projects are following a standardized process for contract administration.
- The vendor provides status reports to the project manager on a routine basis.
- The project manager reports status changes to the vendor on a routine basis.

Contract Administration—Level 4 Characteristics

- Vendors are integrated into the project planning process.
- Contract management is evaluated on a periodic basis.
- Lessons learned and best practices are captured and made available to other projects.

Contract Administration—Level 5 Characteristics

- There is a program in place to continuously improve the contract administration process.
- Lessons learned and best practices are used to improve the contract administration process.

2.3.9.6 Contract Close-Out

Similar to administrative close-out, contract close-out involves assuring that the provisions of the contract have been completely met, that all final reports are completed and filed, and that other closing activities specified in the contract are executed. For a list of survey questions that applies to this process see page 219 in the Appendix.

Contract Close-Out—Level 1 Characteristics

- There is no defined and documented process for contract close-out.
- Project managers follow their own informal process for contract close-out.
- Contract acceptance follows an informal process.

Contract Close-Out—Level 2 Characteristics

- There is a defined and documented process for contract close-out.
- There is a formal acceptance criterion that is part of contract close-out.
- The client participates in the formation of acceptance criterion.

Contract Close-Out—Level 3 Characteristics

- All projects are using a standardized contract close-out process.
- Formal acceptance by the client is communicated to all stakeholders.

Contract Close-Out—Level 4 Characteristics

- Contract close-out is evaluated at the organizational level on a periodic basis.
- Lessons learned and best practices are captured and made available to other projects.

Contract Close-Out—Level 5 Characteristics

- There is a program in place for the continuous improvement of the contract close-out process.
- Lesson learned and best practices are used to improve the contract close-out process.

2.4 Points to Remember

The following is a list of important points to remember from this chapter:

- The SEI CMM® for software engineering defines five maturity levels: initial, repeatable, defined, managed, and optimized.
- The P-CMM® is derived from the SEI CMM®.
- The PMMM is derived from the SEI CMM®.
- PMMM level 1: Everyone does his or her own thing.
- PMMM level 2: There is a structured process in place but its use is voluntary.
- PMMM level 3: Everyone is required to use the structured process.
- PMMM level 4: Project decisions are integrated into business decisions.
- PMMM level 5: There is a continuous improvement process in place.

References

[1] Humphrey, W. S., "Characterizing the Software Process: A Maturity Framework," Software Engineering Institute, CMU/SEI-87-TR-11, DTIC No. ADA182895, June 1987.

[2] Paulk, M. C., et al., "Capability Maturity Model for Software," Software Engineering Institute, CMU/SEI-91-TR-24, DTIG No. ADA240603, August 1991.

[3] Project Management Institute, *A Guide to the Project Management Body of Knowledge, 2000 Edition*, Newtown Square, PA: Project Management Institute, 2000.

3

Assessing and Reporting Maturity Level

At this point we have a good grasp of the CMM® as applied to project management. We have described maturity at all five levels for each of the 39 processes that makes up the nine knowledge areas of the PMBOK standard. We have seen that level 3 is the transition between having a documented process (level 2) and having all project teams using the process (level 3). Level 3 defines project management process practice (PP) level maturity. Level 2 defines project management process definition (PD) level maturity. Levels 1, 4, and 5 define both PD and PP level maturity. In this chapter we turn to the measurement and assessment of project management maturity in the organization. One survey document has been established to measure both PD and PP maturity. In other words, our assessment of level 2 maturity will be totally based on a review of the established and documented processes for completeness. All other maturity levels will assess not only process by further examination of PD and documentation but also by an examination of if and how project teams are using the defined processes. In other words the practice component of maturity is measured by examining ongoing and recently completed projects and how the defined and documented processes were used in the execution of the projects. The survey has been designed to accommodate either PD or PP maturity assessments. The survey is modular. A single process may be assessed for either PD or PP maturity or both. Also, all of the processes that define a specific knowledge area may be similarly assessed.

PD is the documented and standardized processes that drive all project management activity in the organization. It is developed in collaboration with those who are expected to utilize it as well as all those managers who either affect it or are affected by it. The development of PD is only the first step in creating a project management culture in an organization. The second, and more difficult step, is the adoption of the project management processes. This will be measured by PP. Project managers, especially those who have come from other organizations, will bring their own approaches to project management. They will be comfortable using their own processes and will have to be sold on PD as being a better approach. Tracking changing values of PD and PP over time will be an important indicator of the success of any project management improvement efforts. All of the data used in this chapter is real data taken from a recent client engagement.

3.1 Overview of the Survey Questionnaire

The initial version of the survey questionnaire was developed as part of a larger consulting engagement aimed at establishing a continuous quality improvement program for project management at a large retailer. The actual questions developed for that engagement are given in the Appendix.

3.1.1 Design of the Survey

From the outset it was clear that two types of assessments would be needed. The first was an assessment of the maturity of the process based solely on a review of its documentation. Whatever the current state of the process definition, it certainly would not be the complete definition. Parts would be missing and some parts may not be correctly defined or documented and in need of further development. Some parts might be adequately defined but not compatible with the culture of the organization and in need of further development. Still others may be correctly defined and documented but need training support for the teams to use them correctly. The assessment would define exactly where the project management processes were with respect to their level of maturity, and given that there were established goals for maturity, a baseline would be established as well as the gap between target PD maturity and actual PD maturity.

The second was an assessment of how projects were putting the processes into practice. Each project would have its own assessment. Some projects could be expected to operate below the baseline maturity level, others near the baseline, and still others above the baseline. It was clear that the ability to compare process maturity and practice maturity was necessary. To that end, the survey was designed so that we could conduct the two types of assessments with the same survey.

To cover all 39 processes at all five maturity levels required 195 separate maturity assessments, and each maturity assessment would most likely include two or more questions. The total number of questions could easily get out of hand. Even with the efforts to contain that number, the total number needed to adequately cover all 195 categories was 807. To conduct such a survey without overwhelming the person being interviewed led me to structure the sequence of questions in such a way as to minimize the time and effort required of both the interviewee and the interviewer. The questions had to be stated simply and in such a fashion that individual biases would be minimized. I decided to structure each question so that it could be answered either "Yes," "No," or "Not Applicable." Since the attainment of a maturity level was dependent on having attained the immediately preceding maturity level, I decided to sequence the questions within process by maturity level. That meant that if questions at one level were largely answered with a "No" response, the same would likely be the case for questions at all higher maturity levels. That is not to say that all further questions had to have a "No" response, but that the preponderance of questions would have a "No" response. In actual practice, that seemed to hold true and did translate into a more humane treatment of the interviewee. As I observed through the interviews, the sequencing did create a learning curve effect. Interviewees quickly learned the structure and could answer blocks of questions at one time rather than having to answer questions one at a time. The typical interview covered all 807 questions in a little less than 2 hours.

The "Not Applicable" response was used exclusively in the assessment of PP. For example, some projects did not include a procurement phase, and therefore, questions about the procurement management processes were outside the scope of the project. Procurement management may, in fact, be part of the project management processes, but that assessment would be covered by the PD. Once a PD baseline maturity level had been established, further assessments would only have to include those processes that had changed. PD maturity level assessments could also be done on selected processes that had been problematic or otherwise were areas of concern.

3.1.2 Defining Maturity Level Penetration

Simple ratios are used to define maturity level penetration (MLP) at both the process level and at the summary level for each knowledge area. MLP is a numeric value that states the highest maturity level attainment plus the proportion of the next maturity level attained. To attain a specific maturity level, a process must meet all the requirements of that level as well as all previous levels. In other words, maturity is a cumulative measure. The maturity level penetration of process i at maturity level j is calculated using the formula

$$mlp_{ij} = (\text{\# of Yes responses at level } j) \, / \, (\text{\# of Yes responses at level } j + \text{\# of No responses at level } j)$$
$$\text{for } i = 1, 2, \ldots 39, j = 1, 2, \ldots 5$$

Notice that the "Not Applicable" response is not included in the calculation. MLP is a very flexible and useful metric that we will have occasion to use frequently in the remainder of the book. To see exactly how we will use it let us look at an example.

Example. In this example we have calculated the following values for process *i* definition:

$$mlp(pd)_{i1} = 1.0$$

$$mlp(pd)_{i2} = 1.0$$

$$mlp(pd)_{i3} = 1.0$$

$$mlp(pd)_{i4} = 0.2 \qquad \text{implies that } MLP(PD)_i = 3.2$$

$$mlp(pd)_{i5} = 0.1$$

This profile of mlp values tells us that process i meets all of the requirements to be at level 1, level 2, and level 3, and 20% of the level 4 requirements. By definition of maturity levels, we know that a process must meet all requirements of the previous maturity levels to be considered at the next level. For this example, that means that process *i* has attained maturity level 3. Furthermore, it has penetrated 20% into the next maturity level, maturity level 4. Therefore, we define the MLP of process *i* to be 3.2. That is, $MLP(PD)_i = 3.2$.

Continuing the example, suppose we calculated the following values for process *i* practice from a project that has just been completed:

$$mlp(pp)_{i1} = 1.0$$

$$mlp(pp)_{i2} = 0.3 \qquad \text{implies that } MLP(PP)_i = 1.3$$

$$mlp(pp)_{i3} = 0.1$$

$$mlp(pp)_{i4} = 0.2$$

$$mlp(pp)_{i5} = 0.0$$

Because $MLP(PD)_i = 3.2$ and $MLP(PP)_i = 1.3$, the project that gave rise to this data operated below the baseline maturity level for process *i*. This may be an anomaly, in which case the project should be further audited to discover and correct the problem. Other projects may exhibit the same performance in which case the problem may be systemic. In that case further analysis at the PD level is required. The root cause has to be identified and corrected. We will deal with the details in Chapters 5 and 6.

For the sake of another example let us reverse the data so that $MLP(PD)_i =$ 1.3 and $MLP(PP)_i = 3.2$. While this may seem to be an anomaly, it is a possibility. This is an indication that there may be level 3 best practices in the project being assessed that can be used by other projects and used to increase $MLP(PD)$ to 3. This should be further investigated through process improvement initiatives for process i. We will deal with the details in Chapters 5 and 6.

3.2 Reporting the Process Maturity Baseline

All of the maturity data that we will be reporting lends itself to graphical representation. This section and the next introduce, with applications, two graphical tools that are intuitive and insightful of the underlying PD and PP maturity and the relation between them. They are the Kiviatt Charts and Box & Whisker Plots. These tools are flexible and adaptive and can be formatted to allow us to superimpose multiple sets of PD and PP maturity data on the same graphical plot and draw a variety of inferences and conclusions.

3.2.1 Kiviatt Charts

A Kiviatt Chart (source unknown) uses a circle to portray multidimensional data. Each dimension is represented by a line segment radiating from the center of the circle and extending out to the outer edge of the circle. The line segments are equally distributed around the circle. Every line segment has the same scale. Low values are at the center of the circle and high values at the outer edge of the circle. I have used Kiviatt Charts to represent as many as 20 dimensions and have not sacrificed the intuitive properties of this charting process. Kiviatt Charts have been in use for a number of years in a variety of applications. Several examples follow.

Figure 3.1 is an example that shows thinking style data produced by the Herrmann Brain Dominance Instrument (HBDI) [1]. There are four dimensions: analyze, organize, personalize, and strategize. The analyze thinking style is characteristic of an individual who thinks logically, who is analytic, who is generally a technical person, often mathematical, and one who likes to solve problems. The organize thinking style is characteristic of an individual who tends to be administratively focused; they are often conservative and like to follow process and procedure; these individuals lead a controlled work life and like planning and related tasks. The personalize thinking style is characteristic of an individual who is very "touchy-feely." They are interpersonal, emotional, and

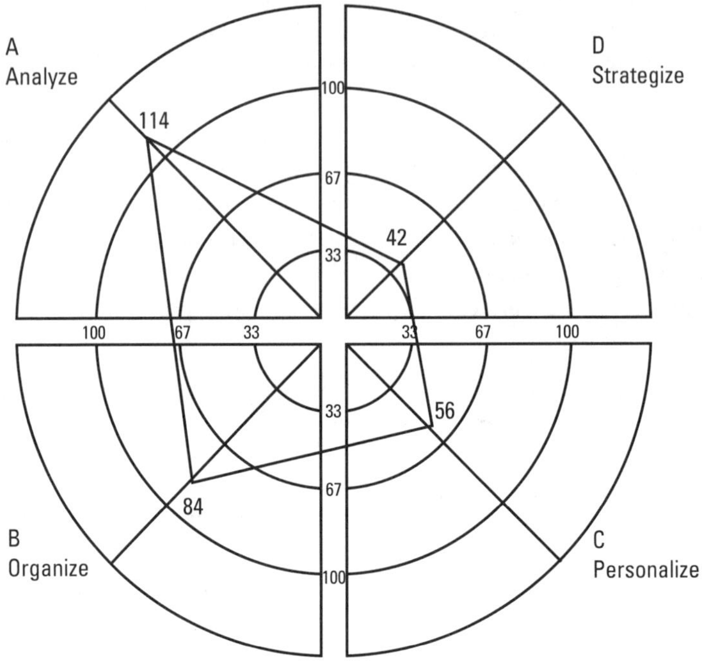

Figure 3.1 Use of the Kiviatt Chart to display HBDI data.

very social, and often they are spiritual and musically inclined. The strategize thinking style is characteristic of an individual who is a conceptualizer, who is imaginative, and one who is particularly good at synthesizing a situation. Each dimension is measured on a scale that ranges from 0 to 133. The data points are determined from a 180-question survey that the individual completes. The responses are analyzed and filtered through an algorithm and four scores are produced. In the example the four scores are 114, 84, 55, and 42, respectively. A closed quadrilateral, called a kite, is formed from the four scores, as shown in Figure 3.1. High scores indicate a preference for that style of thinking. Low scores indicate an avoidance of that thinking style. The example is of an individual who is primarily a left-brain person as indicated by the high scores on the A and B quadrants. The kite can be symmetric, representing those individuals who equally prefer and use all four thinking styles. The kite can be very asymmetric, representing those individuals who have a strong preference for three, two, or one thinking style. A strong preference for a single thinking style and an avoidance of the other three produces the most asymmetric kite.

Figure 3.2 is another example of the Kiviatt Chart. Here the figure displays the skill profile of a project team on five skill groupings: project management, management, business, interpersonal, and personal. Each skill grouping consists of an aggregate score based on a number of individual skill proficiencies.

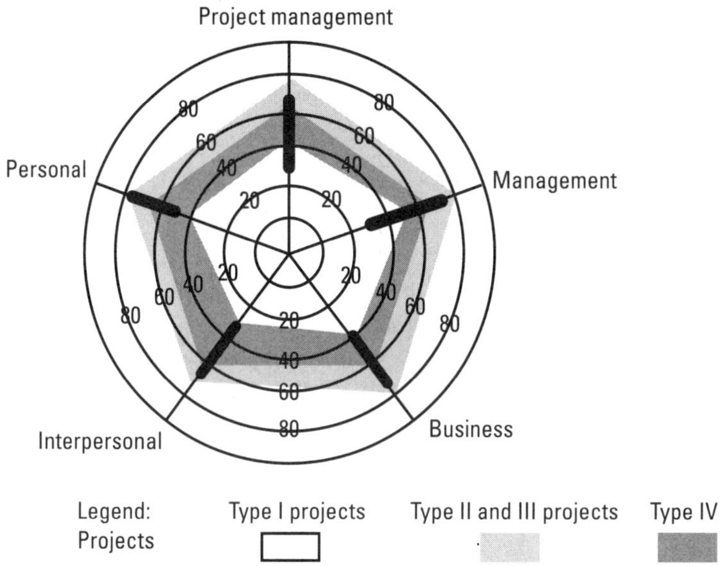

Figure 3.2 Use of the Kiviatt Chart to display a skills profile

The bar represents the range of proficiencies of the team. Superimposed on the Kiviatt Chart are the ranges of skill levels that should be present in a team for a specific type of project. For this example there are four types of projects that range in complexity from critical mission (type I) to simple (type IV). This particular team does not display a skill profile sufficient for type I projects.

Figure 3.2 displays only the range of skill proficiencies. Some prefer to see actual values. Figure 3.3 displays the skill proficiencies of each team member on each of the five skill groups superimposed on one Kiviatt Chart. Knowing the distribution of skill proficiencies within the range does convey more information than the range alone but it is a more complex graphic. I guess you get what you pay for. For even more detail, the skills in a specific skill group could be displayed individually in the Kiviatt Chart format.

Figure 3.4 shows a Kiviatt Chart for nine dimensions. You can see that the figure is still quite intuitive and not at all cluttered even for this many variables. The data here represents the affinity of a project team for each of the nine roles that Belbin's [2] research shows is essential for every team to be effective.

Following are brief definitions of each of the nine roles.

- *Plant:* Someone who is expected to bring creative and new ideas to the team. They often will be instrumental in helping the team solve problems.
- *Resource investigator:* These individuals are good at reaching out for resources and linking the team to external ideas that may be useful.

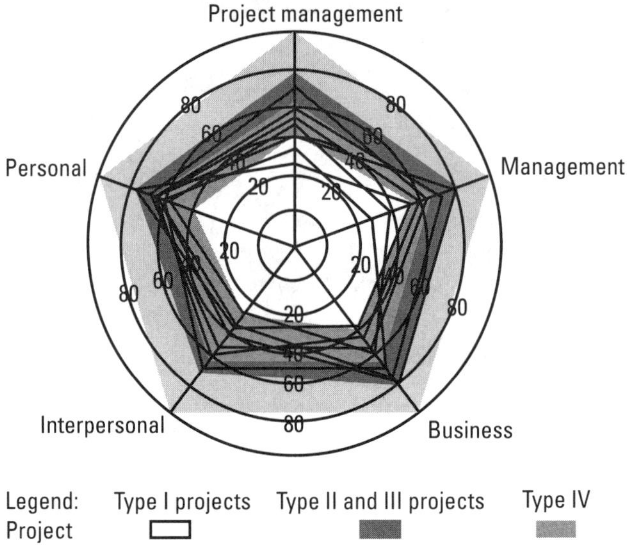

Figure 3.3 Use of the Kiviatt Chart to display several individual skill profiles.

- *Coordinator:* These are the leaders of the team. They take advantage of team strengths and avoid weaknesses and generally make sure that the project plan moves ahead smoothly.
- *Shaper:* These are the planners. They set objectives and establish priorities.

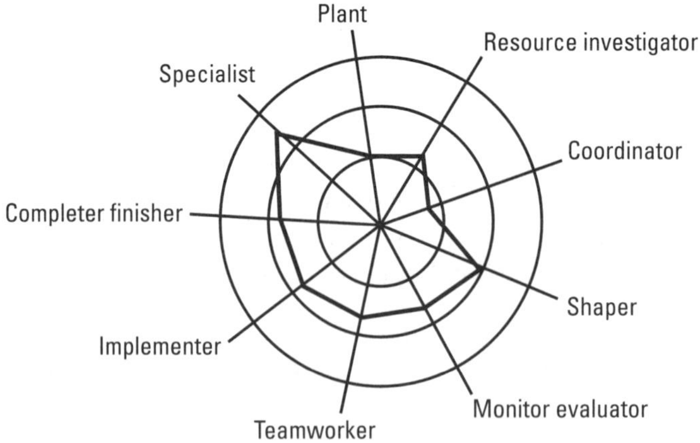

Figure 3.4 Use of the Kiviatt Chart to display Belbin's team role data.

- *Monitor evaluator:* These are the analytical members of the team. They study the team's problems and evaluate alternative solutions. They are a good balancing factor on the team when it comes to decision-making.
- *Teamworker:* These team members have a talent for bringing the team members together into a functioning unit.
- *Implementer:* These team members have a talent for turning the plan into action. They make things happen.
- *Completer/finisher:* These are the detail members of the team. They make sure that things happen as they are supposed to happen.
- *Specialist:* These are the subject matter experts on the team. They are expected to contribute their expertise and that is all.

Higher values are plotted towards the outer edges of the circles. This team has an observed weakness in plant, resource investigator, and coordinator roles. For a discussion of these roles see Belbin [2].

The Kiviatt Chart can also be used to compare two sets of data along the same dimensions. For example, the team's HBDI profile from Figure 3.1 can be superimposed on the project's HBDI profile, producing Figure 3.5. This use of the Kiviatt Chart is unique and provides a powerful tool for the project

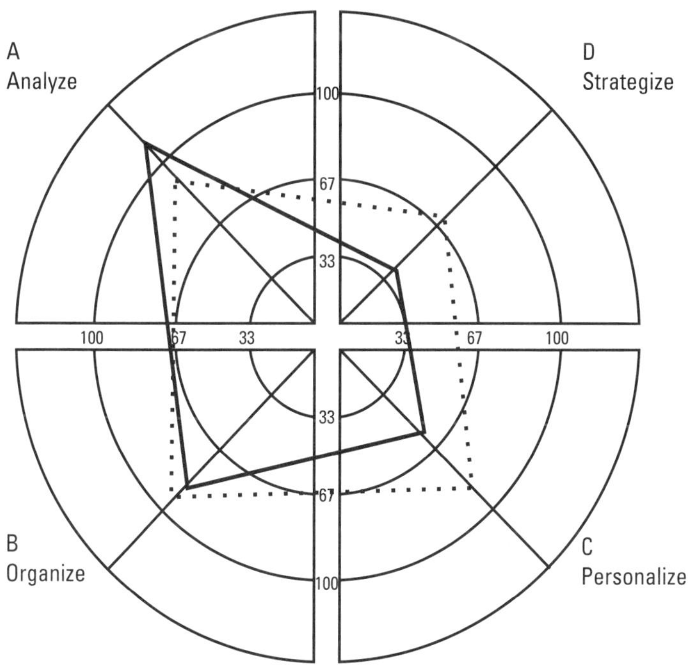

Figure 3.5 Use of the Kiviatt Chart to compare the team and project HBDI profiles.

manager. It shows the degree of alignment between the project and the team using the HBDI as the metric on which that comparison is made. In this example the team has a preference for left-brain thinking. The project, on the other hand, displays more of a balance among all four thinking styles. In this case there is a distinct gap in the personalize and strategize quadrants. The project manager is aware of this misalignment and can put the appropriate corrective measures in place. The interested reader can consult Wysocki [3] for more detail on how to analyze this type of data.

As you can see, the Kiviatt Chart is quite adaptable to a number of situations. We will have many occasions to use it as we examine maturity level data for the 39 project management processes. It will be used for decision making at the process level and at the knowledge area level.

3.2.2 Box & Whisker Plots

Our second graphical tool is the Box & Whisker Plot. The Box Plot was originally formulated by Mary Eleanor Spear [4] and further developed by John Tukey [5] and is the basis for what is now called the Box & Whisker Plot. Figure 3.6 is an example of the graphic icon that is used to display the range, interquartile range, and median of a set of numeric data. (An example of how we will use the Box & Whisker Plot is shown in the next section in Figure 3.8.)

Despite the fact that the Box & Whisker Plot has been around for more than 50 years, its use is not that widespread. First, and most important, is that it is intuitive. It needs little explanation. It is a powerful visual summary of the parameters of a sample set. The size of the box is an indicator of the variance and symmetry in the sample. The location of the median with respect to the interquartile range is another indicator of the symmetry in the data. It will be our favored way of displaying and comparing PD and PP maturity level data.

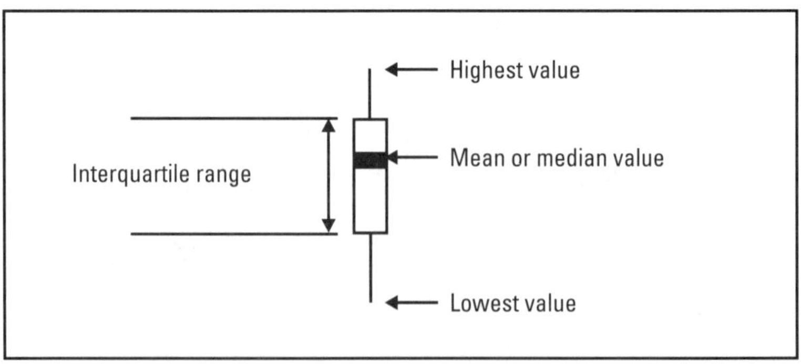

Figure 3.6 The Box & Whisker Plot icon.

3.3 Reporting the Project/Process Maturity Gap

The Kiviatt Chart and Box & Whisker Plots are intuitive and offer great flexibility and economy in displaying maturity level data. PD and PP data can be superimposed to give a complete picture of the project management maturity of an organization's project management infrastructure and project portfolio performance with respect to a documented standard. In this section we begin to explore the variety of situations that our graphic tools can show. In later chapters we will apply these tools to specific improvement initiatives.

Figure 3.7 shows the process level maturity data (PD baseline) with project PP data superimposed. The PP data was collected from nine projects that had just recently completed. Figure 3.8 shows the same data using the Kiviatt Chart format. Figure 3.7 tells us a lot about what is going on with the process definition itself, the process practice, and the relationship between the two. The Plot could also have the goal maturity levels plotted for each process, but that is not done in this example. Note that there are two processes (integration and scope) whose PD values lie significantly above the PP distribution. Most of the projects surveyed for this data set are not using the integration and scope processes as currently defined and documented. For two processes (quality and HR) the PP distributions lie mostly above the PD baseline values. Procurement looks strange. Here most of the PP values are near the PD baseline except for one or two values that are significantly higher. These anomalies should raise our curiosity and suggest further investigation. In my experience with this type of analysis, it is highly suggestive of best practices that were brought to the team by a member new to the organization. This is the diamond in the rough and must be investigated further. The size of the interquartile range also tells us something about the process. Look at time, cost, HR, and risk, for example. The interquartile ranges are small

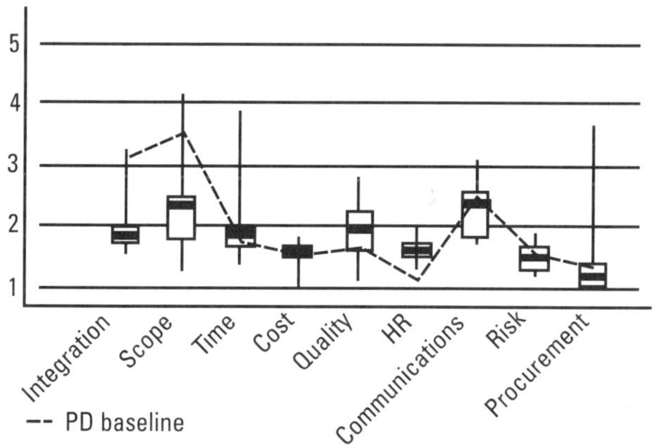

Figure 3.7 Box & Whisker Plot of PD versus PP maturity level data.

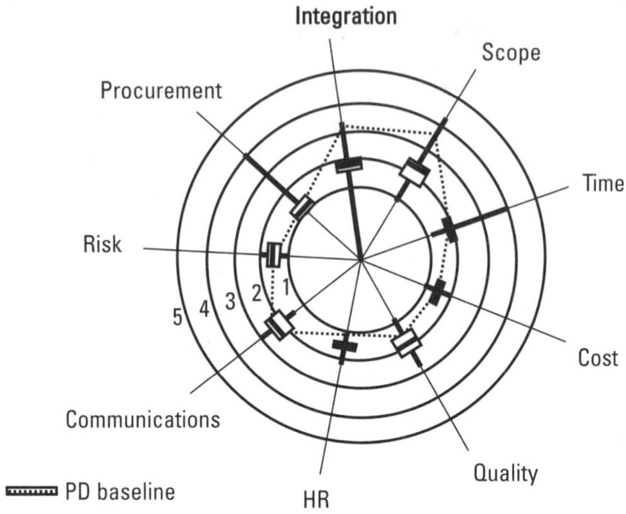

Figure 3.8 Figure 3.7 data shown in Kiviatt Chart format.

and the median value lies very close to the PD maturity level. This is indicative of a process that is well understood and practiced by the teams. Everyone is doing pretty much the same thing and is in alignment with the process maturity. The only issue here is the PD maturity level relative to the target level. Scope, quality, and communications do not display this small variance property.

Figure 3.8 shows the same data as is given in Figure 3.7 but using the Kiviatt Chart format. The Kiviatt Chart format can display more data than the Box & Whisker Plot but is a little less intuitive. Which one you use is really a matter of taste. Both formats are intuitive and convey the same picture. These figures illustrate three situations that are discussed in the following three sections.

For those who would rather see the data at the individual project level, we can replace the Box & Whisker icon with the actual data (Figure 3.9). There is some sacrifice in the intuitiveness from using the Box & Whisker Plot.

Figure 3.10 is yet another example of the use of Box & Whisker Plots. In this example we are tracking changing PD and PP values for a single process. This type of data display is most useful as a summary level report of improvement initiative results for a single process over time. Again, the Box & Whisker Plot tells us quite a bit about what has been going on with this process and the improvement initiatives. For this improvement program both the PD and PP maturity levels are being affected. At the end of each quarter the PD maturity level is assessed and a sample of projects are surveyed with respect to the process under observation. Note what is happening. First, the PD maturity level is increasing toward some target value. The PP maturity level of the sampled projects is moving up slowly. Keep in mind that the project teams are aiming at a

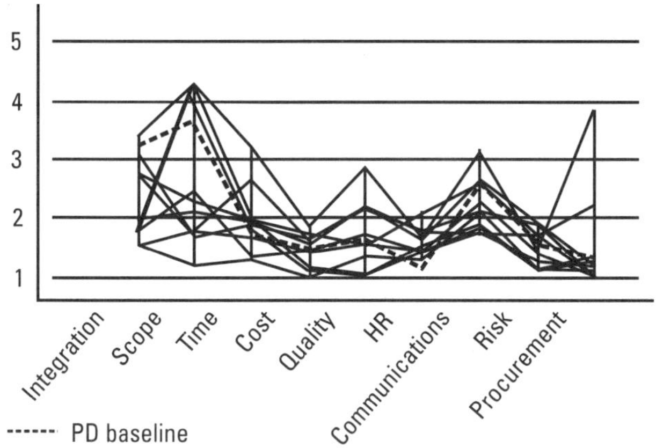

Figure 3.9 Plot of individual project PP maturity values versus PD baseline.

moving target. That translates into considerable change in how projects use the process even as it changes. Note that the PP distributions are converging on the PD values and the interquartile range is decreasing as well. This is suggestive of a good deal of coordination between process and practice as well as an appropriate training program for the new and changed process that is affected. All of this points to a well-executed improvement initiative.

3.3.1 PP Below PD Baseline

In Figure 3.7 note that the maturity level of the process (dashed line) falls above the practice data collected from recently completed projects (the Box &

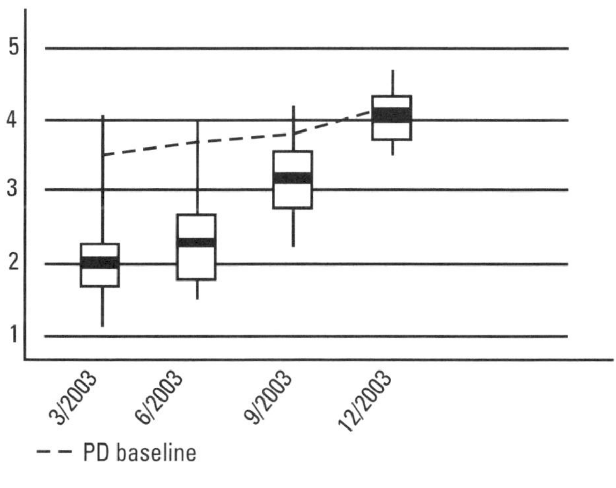

Figure 3.10 Box & Whisker Plot for improvement initiatives for a single process.

Whisker Plot data). I consider situations where the interquartile range lies below the PD maturity level to be significant and worth further investigation. That is the case with the integration and scope knowledge areas.

3.3.2 PP at PD Baseline

If the preponderance of projects are practicing a process at about the same maturity level as the process, that would indicate good alignment between process and practice. Figure 3.7 shows examples of this in the time, cost, risk, and procurement knowledge areas.

Apart from the positioning of the interquartile range of the PP data with respect to the PD value, the variance of the interquartile range is of interest. If the variance is small, it indicates relative uniformity in the practice of a process regardless of its relationship to the PD value. To have the PP interquartile range small and have it include the PD baseline would be an ideal situation. It reflects harmony between defined processes and the practice of those processes. It would indicate a common understanding of the process across all project staff and further that the process definition is aligned with accepted industry practices. If the variance is large, there may be some projects whose PP value is greater than the PD value and worthy of further investigation for potential best practices. This situation is more likely to be the one you encounter. It reflects a range of understandings, acceptance, and agreement across all project staff. As the variance of PP maturity values around the PD baseline increases, there is more reason to be concerned. There should be an investigation to ascertain the reasons for this irregular behavior. There may be behaviors to correct or best practices to incorporate into the project management process. In both cases PP and PD maturity levels will be positively affected.

3.3.3 PP Above PD Baseline

In Figure 3.7 note that the PD maturity level falls below the PP maturity level distribution collected from the quality knowledge area in recently completed projects. I consider situations where a significant number of PP data points lie above the PD maturity level to indicate a serious anomaly.

You might wonder how that can happen. How can projects' PP values lie above the established PD maturity level? This can happen if the development and documentation of project management processes seriously lags accepted industry practices and standards. Project managers are often familiar with common industry practices and will ignore the processes as defined by their organization if they believe something better is available to them.

3.4 Maturity Profile by Knowledge Area

The maturity profile for each of the nine knowledge areas is built from a concatenation of the maturity profiles of each of the processes that define the knowledge area. This process level maturity assessment is a more granular assessment than used in other approaches. The primary reason I have taken this approach is to provide an analysis path for improving the maturity level of a single knowledge area. If the assessment data suggests that a particular knowledge area is performing below nominal, further analysis of the maturity level data for the processes that make up that knowledge area can then be undertaken without any further maturity data collection. That more detailed analysis will certainly isolate areas where maturity performance is below expectations and may shed some light on improvement initiatives that should be undertaken to correct the less than acceptable maturity level performance in the cited knowledge area. Chapter 6 provides considerable detail on the identification and analysis of maturity level problems and the subsequent commissioning of improvement initiatives.

Even though we are going to use the gaps between the existing PD and the target level PD and the gap between the PD maturity values and the PP values as a criterion for selecting improvement opportunities, we need another metric to help prioritize the importance of the project management processes with respect to overall business objectives. It is possible that large gaps may be associated with the processes that are relatively less important than other processes that have smaller gaps. Our approach will combine the relative importance of processes and their PD and PP gaps to prioritize processes for improvement initiatives. Our first order of business is to establish a model for prioritizing the 39 processes. That is the topic of the next section.

3.4.1 Process Maturity Matrix

In a seminal paper published in the *Harvard Business Review* in 1987, Maurice Hardaker and Bryan K. Ward [6] introduced process quality management (PQM). The purpose of the PQM was to give management an organized approach to improving the performance of the business. It would do that by identifying and prioritizing the processes of the business that would have maximum impact on the critical success factors that drive the performance of the business if the processes associated with them were improved.

The heart of PQM is a matrix whose rows are the business processes of an organization and whose columns are the agreed upon critical success factors (CSF) of the organization. In our use of the PQM the rows will be the 39 project management processes. The columns will be the CSFs of the project management methodology. Two additional columns are appended to complete this matrix. The first is a count of the number of CSFs that are related to each of the business processes. For our purposes we call these numbers the correlation factors. They range from 0 to the number of CSFs. The second is an evaluation of the quality of each business process. Hardaker and Ward measure quality as follows: A for excellent, B for good, C for fair, D for bad, and E for embryonic. For our purposes I have modified that column to be the MLP value of the process.

Obviously the PQM is robust; that is, it has wide applications to areas other than business processes and CSFs. For example, I have used it in the health care industry by substituting direct patient care services for business processes and quality performance metrics for CSFs. The analysis is identical even with these changes. For the purposes of this book I have replaced business processes with the 39 project management processes, CSFs with the top 10 reasons why projects succeed as reported by the Standish Group [7], and the evaluation of quality with project management maturity level. The resulting process maturity matrix is shown in Figure 3.11, and the process maturity zone map is shown in Figure 3.12. The maturity level data is from a recent consulting engagement with a large retailer.

As indicated in Figure 3.11 the PQM consists of three parts: a matrix defined by the 39 project management processes and the 10 project CSFs, correlation factors ($c_{i\cdot}$, i = 1, 39), and maturity level assessments ($MLP\,(PD)_i$, i = 1, 39).

$$C_{ij} = \begin{cases} 1, \text{if the attainment of CSF } j \text{ is dependent upon process } i \\ 0, \text{otherwise} \\ c_{i.} = \sum_j c_{ij} \end{cases}$$

The last two columns provide the data that defines another matrix called the zone map. That data is graphically summarized into a matrix called the zone map, as shown in Figure 3.12. Processes whose data points lie towards the northwest part of the zone map are processes that should be considered for any process improvement initiatives because they have a low maturity level and are associated with several CSFs. The zone map gives us a crude tool for prioritizing processes for improvement initiatives. In the example, data processes P21, P14, P23, P28, and P33 would receive immediate attention. As improvement initiatives are completed, one would expect to see these processes move from left to right in an updated zone map. Once some of those improvements are in place, a

Knowledge area	PM process	CSF1	CSF2	CSF3	CSF4	CSF5
Integration mgt	Project plan dev.	1	1	1	0	1
	Project plan execution	0	1	1	0	1
	Int. change cont.	1	1	1	1	1
Scope mgt	Initiation	1	0	1	1	0
	Scope planning	0	1	1	1	1
	Scope definition	0	1	1	1	1
	Scope verification	1	1	1	1	1
	Scope change cont.	1	1	1	1	1
Time mgt	Activity definition	0	0	1	0	0
	Activity sequencing	0	0	1	0	0
	Activity duration est.	0	0	1	0	0
	Schedule dev.	0	1	1	0	1
	Schedule control	1	1	1	0	1
Cost mgt	Resource planning	0	1	1	0	0
	Cost estimating	1	0	1	0	0
	Cost budgeting	0	0	1	0	0
	Cost control	1	0	1	0	1
Quality mgt	Quality planning	0	0	1	0	0
	Quality assurance	0	1	1	0	1
	Quality control	1	1	1	0	1
HR mgt	Org. planning	1	1	1	0	0
	Staff acquisition	1	0	1	0	0
	Team development	1	1	1	0	0
Comm. mgt	Comm. planning	1	0	1	0	0
	Info. dist.	1	1	1	1	0
	Perf. reporting	1	1	1	0	0
	Admin. closure	1	0	1	0	0
Risk mgt	Risk mgt planning	0	1	1	0	1
	Risk identification	0	0	1	0	0
	Qual. risk analysis	0	0	1	0	0
	Quant. risk analysis	0	0	1	0	0
	Risk resp. planning	0	1	1	0	0
	Risk monitoring and cont.	1	1	1	0	1
Proc. mgt	Procurement planning	0	0	1	0	0
	Solicitation planning	1	0	1	0	0
	Solicitation	1	1	1	0	0
	Source selection	1	0	1	0	0
	Contract admin.	0	0	1	0	0
	Contract close-out	0	0	1	0	0

Figure 3.11(a) The process maturity matrix.

CSF6	CSF7	CSF8	CSF9	CS10	$C_{i.}$	MLP_i	PM process
0	0	1	1	1	7	3	Project plan dev.
0	0	1	0	0	4	3.33	Project plan execution
1	1	1	0	0	8	3	Int. change cont.
0	1	1	1	0	6	4	Initiation
0	1	1	1	1	8	4	Scope planning
0	1	1	1	0	7	3	Scope definition
0	1	1	1	0	8	3	Scope verification
1	1	1	1	0	9	3.67	Scope change cont.
0	0	1	1	1	4	1.82	Activity definition
0	0	1	0	1	3	2	Activity sequencing
0	0	1	1	1	4	1	Activity duration est.
0	0	1	1	1	6	1.59	Schedule dev.
0	0	1	0	0	5	2.1	Schedule control
1	1	1	1	1	7	1.5	Resource planning
0	0	1	1	0	4	1.36	Cost estimating
0	0	1	0	0	2	2	Cost budgeting
0	0	1	0	0	4	1.22	Cost control
1	0	1	0	1	4	1	Quality planning
0	0	1	1	0	5	1.14	Quality assurance
1	0	1	0	0	6	2.63	Quality control
1	1	1	1	1	8	1.11	Org. planning
0	0	1	1	1	5	1.22	Staff acquisition
1	0	1	1	1	7	1.1	Team development
1	0	1	0	0	4	2	Comm. planning
0	0	1	0	0	5	3	Info. dist.
0	0	1	0	0	4	2.14	Perf. reporting
0	0	1	1	0	4	3	Admin. closure
1	0	1	1	1	7	1.33	Risk mgt planning
0	1	1	1	0	4	1	Risk identification
0	0	1	1	0	3	2	Qual. risk analysis
0	0	1	1	0	3	2	Quant. risk analysis
0	0	1	1	0	4	1.33	Risk resp. planning
1	0	1	1	0	7	1.67	Risk monitoring and cont.
1	0	1	1	1	5	1.71	Procurement planning
0	0	1	1	1	5	1	Solicitation planning
0	0	1	1	1	6	1.29	Solicitation
0	0	1	0	0	3	2	Source selection
0	0	1	1	0	3	1	Contract admin.
0	0	1	0	0	2	1	Contract close-out

Figure 3.11(b).

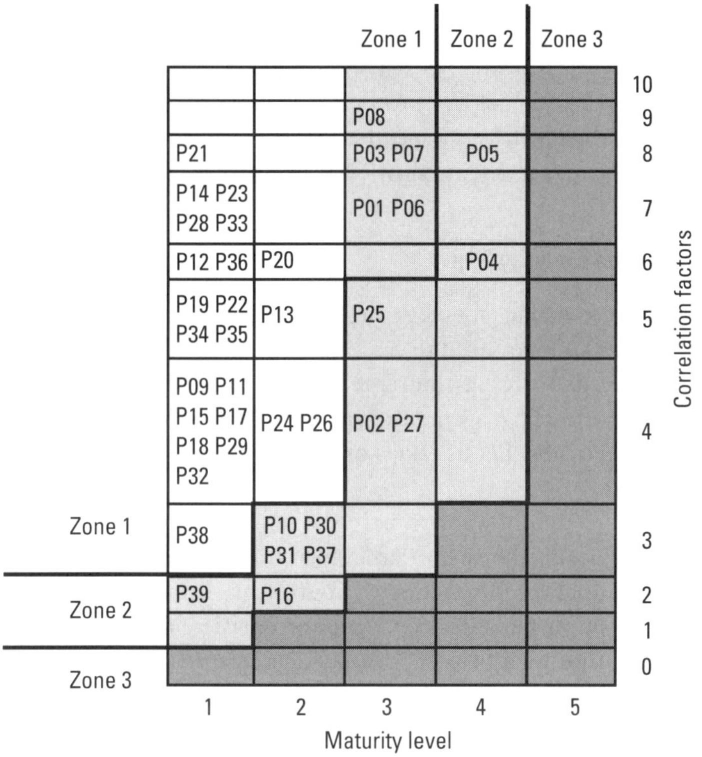

Figure 3.12 The PD maturity zone map.

reprioritization can be done and another round of improvement initiatives undertaken.

If the Box & Whisker Plot report given in Figure 3.7 is used to prioritize knowledge areas for improvement initiatives, the following might happen. The time management knowledge area might be chosen as the highest priority area for improvement. Time management consists of five processes:

1. P09: activity definition;
2. P10: activity sequencing;
3. P11: activity duration estimating;
4. P12: schedule development;
5. P13: schedule control.

Note the location of these five processes in the zone map in Figure 3.12. The prioritization of these five processes for improvement might be P12, P09, P11, P13, and finally P10. P12, P09, and P10 improvement initiatives will be to complete the definition and documentation of the processes. That will increase the PD maturity levels from 1 to 2. Following that the next set of improvement initiatives will be to raise all five processes PD and PP maturity levels to level 3. In other words, there will be a concerted effort to improve the PP maturity level of the entire time management knowledge area to level 3.

3.4.2 Closing the Maturity Gap

Not every project management process needs to be at level 5 maturity, and maybe even none of them need to reach level 5. Also, not every project management process needs to reach the same level of maturity. Setting the maturity goals for your project management processes is more involved than you might have originally thought. Let us take a brief look at some of the factors that need to be considered.

While you certainly want to have the best project management processes and practices you can, the cost of attaining and maintaining them may not be practical. Given that level 5 maturity is out of reach for all 39 processes, what compromise position makes sense? I cannot envision any organization that is serous about cultivating a project management culture that would settle for less than maturity level 3 for all 39 processes. That means that whatever your PD maturity levels might be, you would want your PP values to cluster closely around 3.0. That goal is a difficult one to achieve. Stop and think about the infrastructure you will need to certify that you are operating at level 3 maturity. Every project must undergo a series of project reviews at key points and milestones along the project life cycle. Any anomalies must have a corrective action plan developed, put in place, and monitored for compliance. Process documentation must be complete, clear, readily available, and backed up with the needed training and consulting support. And then you must staff for delivery of all these support services.

The decision to set the target maturity level is not to be taken lightly. To have your project management methodology at level 5 for all 39 processes is certainly an admirable accomplishment but it might be overkill. To achieve that level requires the full cooperation and support of the entire organization because project management must be fully integrated into the culture and operation of the organization. Let us take a quick look at each maturity level and what might be involved.

3.4.2.1 Getting to Level 1: Initial

Since level 1 maturity in project management is characterized by project teams acting in an ad hoc manner without the benefit of any documented processes, the only requirement for level 1 maturity is that the project team is following some approach of their choosing. The approach need not be defined beforehand

and in fact will most likely arise on an as-needed and when-needed basis. There is no stipulation as to how complete or how well thought that approach might be. It might be very sound and based on good experiences from earlier projects of one or more team members or may be process information taken from team member experiences from previous employers. Level 1 maturity is almost automatically attained. The only requirement is that teams be engaged in some form of project management with no specification as to how, when, or why.

3.4.2.2 Getting to Level 2: Structured

Level 2 maturity is a process-oriented versus practice-oriented level of maturity. To be at level 2 means that an organization has undertaken and completed a project to establish a documented and standardized project management methodology. This is a big step for most organizations because it is a precursor to a cultural shift wherein the organization will be asked to integrate these new project management processes and procedures into its other business processes.

3.4.2.3 Getting to Level 3: Institutionalized

The major focus of level 3 maturity is on the utilization of the processes defined as part of attaining level 2 maturity. Getting to level 3 is a daunting task in most organizations. There are several reasons for this:

- The project management process was not developed with broad involvement of those impacted by or expected to use these new processes. Now they do not feel any ownership and hesitate or even push back when expected to use the new processes.
- The project management process(es) do not meet the needs of many of the projects and therefore are not used or used only selectively.
- The project management process documentation is not clear, there are few templates to use, or the templates are too cumbersome. Project teams tend to fall back on those processes and templates they have used successfully in the past or have developed out of necessity.
- The training has been less than enthusiastically accepted and has not been seen as effective.

If you have a role to play in bringing your organization to level 2 project management maturity, you need to develop plans to mitigate these risks. Let us take a look at each one in a little more detail.

There is a prime directive that I will share with you. It is based on many years experience in designing and deploying methodologies. My directive is this: *Whenever you are charged with the responsibility of developing a new process, you must meaningfully involve all of the parties that will affect or can be affected by the new process.* This is accomplished in at least three ways. The first is to populate your task force with a range of professionals who will be expected to use the new process. Look for those whose opinions are highly regarded. If they are respected by their peers, that respect will positively impact the acceptance of their output. This will create ownership and a sense of commitment to the changes about to come. Implementation success is much more probable. The second is to conduct a stakeholder analysis as part of planning the project to develop the new process. That analysis will identify everyone who has a need to know, what they need to know, how they need to know it, and the frequency with which they get to know it. Build a communications management plan into the WBS and execute it! The third is to request and recruit as reviewers those who have a vested interest in the outcome of the process design effort. Give them a chance to comment and critique working drafts and make sure they know you have considered their input.

3.4.2.4 Getting to Level 4: Managed

The step from level 3 to level 4 is a major undertaking. The project management processes are about to become pervasive through the organization. Hardly any business process will be untouched. It is time to integrate the project management process into other business processes and decision making activities. The impact can be at the strategic level (project selection and portfolio management), tactical level (integrating with management decision making and resource allocation), and at the operational level (integrating with other business processes and procedures). In some cases the processes will integrate without much incident, in others some business process redesign may be required, while in still others a complete redesign effort will have to be undertaken.

3.4.2.5 Getting to Level 5: Optimizing

This is the final step and it may be the easiest of all. Level 5 requires that continuous process improvement processes be in place. These will largely involve feedback loops from the practice of project management into an organized program to analyze the feedback, identify process improvement opportunities, and initiate those efforts. Chapters 4 through 8 discuss the entire life cycle of an improvement program.

3.5 Points to Remember

The following is a list of points to remember from this chapter:

- A single survey has been designed to assess both PD and PP maturity levels.

- The Kiviatt Chart can display multidimensional data on a two-dimensionsal graph.
- The Box & Whisker Plot is an intuitive graphic icon that summarizes the distribution of a sample of data.
- PD Baseline and a sample of PP data can be superimposed using a Box & Whisker Plot format to visually show gaps between the process definition and its practice.
- Best practices can be isolated using a Box & Whisker Plot format with a superimposed PD baseline.
- Maturity profiles can be plotted at the process level or aggregated to the knowledge area level.
- The process maturity matrix is used to identify those processes that are highly correlated with project CSFs.
- The zone map can be used to prioritize processes for improvement initiatives.
- Target maturity levels should be set for each process.

References

[1] Herrmann, N., *The Whole Brain Business Book*, New York: McGraw-Hill, 1996.

[2] Belbin, R. M., *Team Roles at Work*, Boston: Butterworth-Heinemann, 1993.

[3] Wysocki, R. K., *Building Effective Project Teams*, New York: John Wiley & Sons, Inc., 2002.

[4] Spear, M. E., *Charting Statistics*, New York: John Wiley & Sons, Inc., 1952.

[5] Tukey, J. W., *Exploratory Data Analysis*, Reading, MA: Addison-Wesley Publishers, 1977.

[6] Hardaker, M., and B. K. Ward, "How To Make a Team Work," *Harvard Business Review*, Vol. 65, No. 6, November–December 1987, pp. 112–119.

[7] Standish Group, *Chaos Chronicles, Version 2.0*, West Yarmouth, MA, 2001.

4

Metrics to Identify Project Improvement Opportunities

In the last chapter we discussed Kiviatt Charts and Box & Whisker Plots. These are two tools that can be used to display PD and PP data in our search for process-wide and practice-wide improvement opportunities. This would typically be the starting point for identifying major areas where improvement opportunities should be focused. In this chapter we will peel back the onion and introduce three additional tools that can be used at the individual project level to identify individual projects whose performance suggests that there is a potential improvement opportunity, first in the performance of the project itself, and second in the way the project is using project management processes.

4.1 Project Level

The three tools introduced in this section apply to individual projects. The first two tools—cost schedule control and milestone trend charts—provide high level summary data on the cumulative progress of the project against its plan. Projects whose performance is at variance with its plan are projects that should be further investigated to identify and correct the aberrant performance. The third tool—project reviews—consist of status presentations by the project manager to a panel of experts. The purpose of these reviews is to understand the status of the project and review the project manager's corrective action plans or suggest corrective action plans to the project manager.

4.1.1 Cost/Schedule Control

Cost/schedule control (C/SC), which is also referred to as earned value analysis (EVA), was originally developed by the Department of Defense in 1963 as a tool to monitor the general health of government contracts. The expenditure of money over time compared against the progress over time of the project is combined to provide that monitoring tool. All projects could be reduced to a common measure, and therefore, the status of a portfolio of projects, such as a program, could also be monitored. Let us take a closer look at how C/SC is structured and how it can be used to monitor the general health of a project.

4.1.1.1 Cost Variance

At any point in time a project's general budgetary health can be measured by the conformance of the project to its budget by comparing the planned budget to date against the actual expenditures to date. Two variances can result. If the planned expenditures exceed the actual expenditures, the project budget has been underspent. This can indicate that economies were found or work that was scheduled to be completed was not completed and hence the budget for that undone work was not spent. We cannot tell from this data which situation created the variance. If the planned expenditures are less than the actual expenditures, the budget has been overspent. There are two reasons why this may have occurred. Either the work that was planned cost more than had been estimated, or more work than was planned was completed during this report period and the cost of that extra work put the project over budget. From the budget data alone we cannot determine which of these situations occurred.

4.1.1.2 Schedule Variance

At any point in time a project's general schedule health can be measured by comparing the planned schedule to date against the actual schedule to date. If the actual schedule is ahead of the planned schedule, there are two possible explanations. Either the project manager has found some ways to shorten the work that was to be done or some work that was scheduled to be complete by this time was not completed. The schedule data by itself does not tell us which situation occurred. If the actual schedule has slipped from the planned schedule, there are two possible explanations. Either the work that was planned took longer than estimated or work that was planned for a later date was completed early. From the schedule data itself we cannot determine which of these situations prevailed.

4.1.1.3 Cost/Schedule Interactions

It should be obvious from the discussion of the previous two sections that budget data by itself or schedule data by itself cannot tell the complete story on the general health of the project. The two metrics must be combined. Figure 4.1 is a graphic display that combines cost and schedule data and that tells the whole story.

The following metrics are measured on the report date or update date, which is shown as the vertical line in Figure 4.1. The planned value (PV) (also known as budgeted cost of work scheduled) is the cumulative curve of the planned progress for the project or some significant part of the project. Progress can be measured in terms of planned expenditures or planned work days. Either metric is a good surrogate for progress. The actual cost (AC) curve (also known as actual cost of work performed) measures what was actually expended over the cumulative life of the project as of the report date. The third curve, the earned value (EV) (also known as the budgeted cost of work performed) measures the value (in terms of planned cost or labor) that accrues to the project for completing the work that was completed as of the report date.

Perhaps the most discussed aspect of C/SC is how to measure value. There are basically four approaches. First, credit 100% of the value (budgeted cost not actual cost) when the task is first opened for work; second, credit 100% of the value when the task is 100% complete; and third, credit 50% when the task is first opened for work and the remaining 50% when the task is complete. The fourth method, which is the one I prefer to use, is based on the number of subtasks that make up the task. The EV is based on the proportion of subtasks complete on the report date. Multiply that proportion by the budgeted cost of the task and that is the accrued value for that task. Whichever method you use, the important thing is that EV is clearly defined—there can be no argument about it.

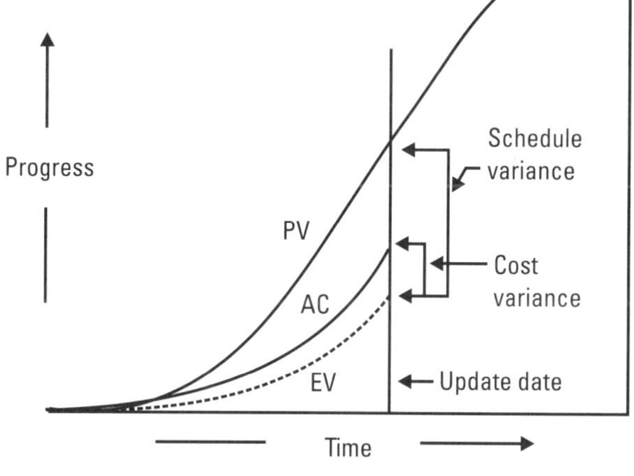

Figure 4.1 The whole cost/schedule control story.

It may be subjective and it may not be theoretically sound, but it is consistent across the life of the project and that is what counts.

Let us get back to the example. In the example the work performed was less than the work scheduled, giving rise to the schedule variance. The budgeted cost of the work performed was less than the actual cost of the work performed, giving rise to the cost variance. In other words the project is over budget and behind schedule—the worst of all possible cases. Less work than was planned was actually accomplished and what was accomplished cost more than was budgeted.

4.1.1.4 Cost Performance Index

There are two ratios that can be calculated from the three metrics discussed above. The first is the cost performance index (CPI), which is defined by the ratio EV/AC. This ratio is a measure of how close the spending pattern is to the planned spending pattern. Values less than one indicate that you are spending more than was planned. Values greater than one indicate that you are spending less than was planned. This ratio can be tracked over time to detect any trends that might suggest either best practices or potential problems for the project being tracked. Whichever is the case, a follow-up action plan should be formulated. CPI history is valuable input to the project review.

4.1.1.5 Schedule Performance Index

The second ratio is the schedule performance index (SPI), which is defined by the ration EV/PV. This ratio is a measure of how close the project is to performing work as it was planned to occur. Values of SPI greater than one indicate an ahead-of-schedule situation. Values less than one are indicative of a project that is behind schedule. These values can also be tracked over time to detect any trends that might suggest best practices or potential problems for the project being tracked. Whichever is the case, a follow-up action plan should be formulated. SPI history is a valuable input to the project review.

Figure 4.2 gives an example of a situation where both performance indexes are positive. This project has accomplished more than was planned to be accomplished at this point in time and did it with less expenditures than were planned. Ahead of schedule and under budget—this is the best of all possible worlds. When a project displays this type of performance, it is important to our improvement initiatives that we investigate the reason for the exemplary performance. The project review may bring those practices to light but a more investigative process might be needed as well. There may be some best practices hidden beneath the covers of this project. We would want to find out what they are and share them with other projects to the extent possible.

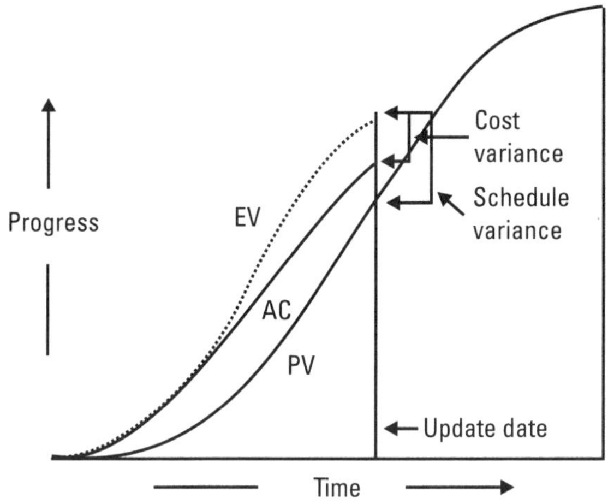

Figure 4.2 An example of positive performance indexes.

4.1.2 Milestone Trend Charts

Another metric for tracking the project schedule over time is the Milestone Trend Chart (MTC). The MTC was introduced in my book *Effective Project Management* [1], which can be referred to for a more complete discussion. Each trend chart tracks a single future milestone event in the project. At each reporting period (weekly or monthly) an updated estimate of the date on which the future milestone event will occur is made. The estimate should be generated by the software package you are using to track the project. Once the work completed since the last report date and the revised estimates to completion are input to the project file, new estimated completion dates of all future tasks are calculated by the software package. The trend in these forecasted dates over time is predictive of the general health and performance of the project. Several milestone events may be tracked for a single project. One milestone event that will be particularly interesting is the completion date of the project. Other milestone events of interest might be the approval of significant phases in the project, such as design or subsystem testing. In the following sections we take a look at several examples and ask the question: "What can we learn about PP based on the project's milestone trends?"

4.1.2.1 Schedule Variance Trends

The MTC focuses on schedule and schedule variances over time. First let us understand how they are constructed and then give a few examples. Figure 4.3 is a typical MTC.

Figure 4.3 shows the forecasted schedule of a single milestone. It may or may not represent a critical path event. One important milestone that the

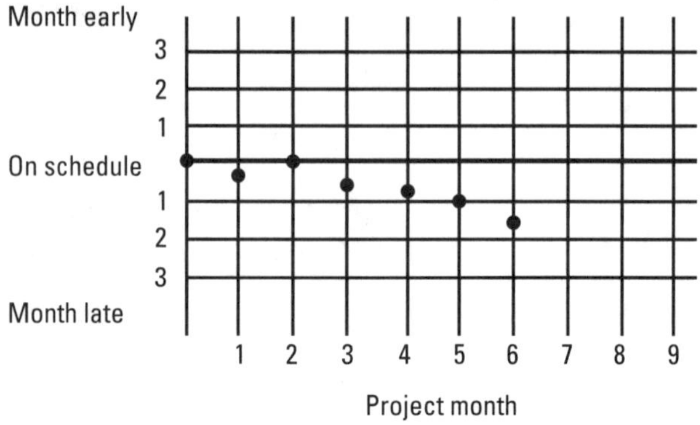

Figure 4.3 A typical MTC.

project manager may wish to track is the project completion milestone. In any case, the milestone trend chart is laid out on a monthly scale with month nine being the planned date on which this milestone event is to occur. Both scales are measured in months. Weekly scales can be used if the situation and project duration suggests that a weekly report frequency would be informative. As project length decreases, a weekly time scale is more likely, but be careful not to overload the team with reports that are not very informative. At each project report date the project administrator can process the project updates through the software package based on the progress of the tasks that were open for work and read off the new estimated date of the milestone event. This data point can be manually added to the MTC to update it to the current report date.

The MTC is a manual report. It can be automated, but that is overkill. After a few months of report data on a milestone event, there will be enough historic information to begin to see trends in the forecasted date of the milestone event. The trends that will be particularly important are those that suggest a growing slippage (indicative of a systemic problem with the project plan or execution) or those that suggest a well-established plan (indicative of best practices). The triggers that we will use to determine whether or not further attention is required are similar to those established in the control charts used in various quality control procedures.

The MTC is a forecasting report. Of all the reports that can be generated for the project, the MTC may be the only one that attempts to forecast the future of a project. It does that by projecting trends into the future. All other project management reports that I am familiar with are reports of historical events. C/SC reports can be used for forecasting, but that is seldom done. The MTC is therefore a valuable report not only for the project but also as input to improvement initiatives.

Now we can interpret Figure 4.3. In month zero, before the project has started, the milestone event of interest is on schedule. No work has been done and so this data point simply reflects the project plan. The month-one status report shows that the forecasted date for this milestone has slipped to 1 week late. How is this to be interpreted? Envision the critical path leading up to the milestone event. Some of the tasks on that critical path were open for work in month one and something happened to those tasks such that their status showed them to be running 1 week later than planned. Because that task or tasks are on the critical path of the subject milestone event, that pushes the updated date of the subject milestone event out 1 week. At the end of month two some of the tasks worked on during that month came in ahead of schedule and corrected the milestone event's forecasted completion date to put it back on schedule. Note that the milestone completion date reflects the cumulative impact of all previous work on tasks that were on the critical path leading to this milestone event. That critical path can change and probably will as project work commences. In the next 4 months we see that there were several slippages that pushed the subject milestone event date out 2 weeks late, 3 weeks late, 4 weeks late, and finally 6 weeks late. Four consecutive data points all trending in the same direction is the trigger that suggests further investigation as to the cause of these slippages. What is happening with this project? Obviously, tasks worked on in each month that were on the critical path leading to the subject milestone event repeatedly came in late. In each report period the tasks that were open for work are not the same tasks that were open for work in the previous report period. Obviously, these different sets of tasks have something in common that has caused these repeated slippages. That certainly indicates a systemic problem with the project plan. Most likely it is a problem resulting from too ambitious estimating of task duration. It might be indicative of a shortage of resources across the entire project or the substitution of less skilled resources. Whatever the reason, it needs to be investigated. There may be some lessons learned here that can benefit this project as well as other projects. The trigger pattern here is four consecutive data points all trending in the same direction. In this case it was 4 months of growing schedule slippages for the same milestone event.

If we change the example to one where the trend is positive rather than negative, what does this suggest? Remember that each report period includes a different set of tasks open for work and the schedule update reflects how those tasks have been handled. In this case the schedule continues to move further ahead. This means that the new set of tasks were completed ahead of schedule as has been the case in the previous 3 months. That suggests duration estimates that were too conservative. It is very unlikely that the trend indicates exemplary project management. Being ahead of schedule for 4 consecutive months does not suggest that the project will remain ahead of schedule. To do so requires the rescheduling of the resources allocated to the tasks that lie further out in the critical path leading to this milestone event.

Figure 4.4 shows a different situation. Here the forecasted milestone date is ahead of schedule beginning in month one and stays ahead of schedule for the next 6 months. Apart from any resource rescheduling obstacles, this milestone event will at least come in on schedule and maybe even ahead of schedule. What else does this MTC tell us about the project? Overestimating task duration is certainly one explanation, but there are more. Except for the aberration in month one, project performance along this milestone's critical path is going pretty much according to plan. That is a good indicator that the entire project plan is well constructed, and I would say that the plan is pretty sound. We should want to know why. What can we learn from this project that may have application in other projects? The trigger pattern here is 7 consecutive months all on the same side of the on-schedule line. In this case those seven data points lie above the on-schedule line. This project was well planned. There may be some best practices hidden in the planning process and just waiting to be discovered.

If the seven data points fell below the on-schedule line, we would have a different interpretation. The project plan would appear to be solid and we would have confidence that the remaining tasks would also be completed as planned. The difference between this pattern and the ahead of schedule variation is that the project has been behind schedule for several months and the project manager has not been able to correct the schedule slippage. That is a serious problem that needs further investigation. Until proven otherwise the project manager is on the carpet.

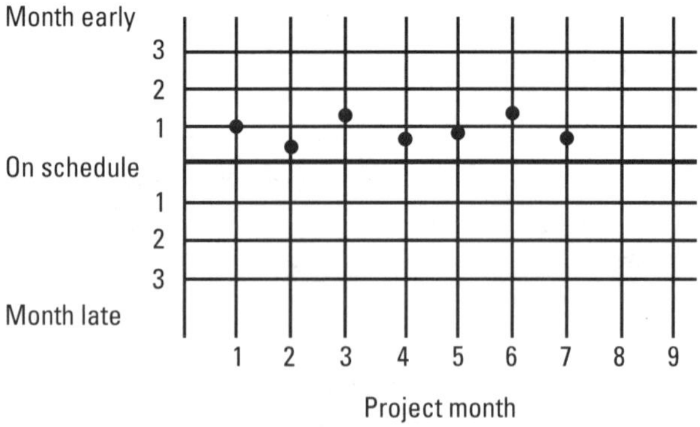

Figure 4.4 A milestone that is consistently ahead of schedule.

4.1.2.2 Performance Index Trends

We can adapt the MTC and use it to plot the CPI and SPI of a single project over time. The horizontal scale stretches out to the planned completion date of the project. The scale can be either weekly or monthly. Figure 4.5 gives an example.

This example tracks both performance indexes on a monthly basis. The most desirable situation is to have both values near 1.0—the on-plan line. Note that the actual cumulative project costs are running slightly below budget at the month six report but that the project is behind schedule. Perhaps the budget situation is explained by the fact that planned work has not been done. That could account for both the positive budget variance and the negative schedule variance. These types of plots are a different way of displaying the information shown in Figure 4.2 and may be easier for senior managers to interpret.

Four different patterns are possible. First, both indexes can be above 1. This indicates exemplary project performance. More work that has been planned at this point in time has been completed and budget expenditures are below the planned level. This is truly a remarkable feat and the reasons should be investigated. Second, both indexes can be below 1. This is the worst of the four patterns. Being behind schedule means that some of the planned budget would not have been spent but yet the project is over budget. Again, the reasons should be investigated. Third, CPI can be above 1 and SPI below 1. This suggests that the work that has not been done has not incurred the planned expense. The project may or may not have spent what was budgeted for the work that has been done. You must investigate and find the cause. Fourth, CPI can be below 1 and SPI above 1. This is the mirror image of the third pattern. Here the over-budget expenditure may be due to the fact that more work was done during the report period than was planned. You must investigate and find the cause.

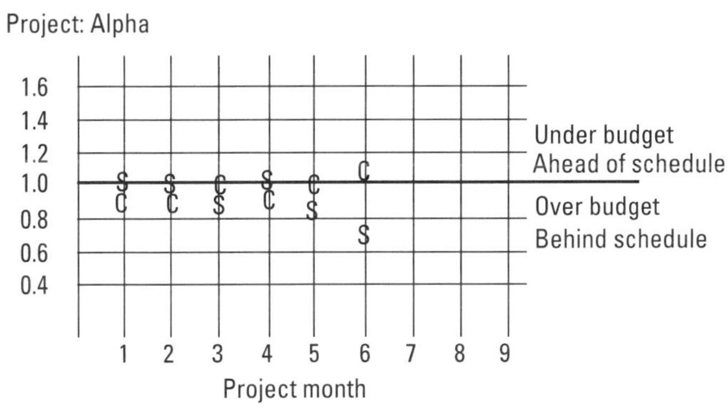

Figure 4.5 Using milestone trend charts to plot project performance indexes.

In all four of the above situations there can be hidden gems just waiting to be discovered. If any of the resulting investigations uncover best practices, these need to be documented, archived, and shared with other projects.

4.1.3 Project Reviews

The project review is a formal means of assessing the general state of health of a project and of discovering best practices and lessons learned for use in other projects.

4.1.3.1 Purpose

There are a number of reasons why an organization might want to conduct a project review. In general, the purpose of a project review is to assess project performance, offer suggestions to the project manager, review and approve corrective action plans, learn about particular nuances that the project manager and team may be using, assess how the project team is using established standards and templates, and identify best practices and lessons learned.

4.1.3.2 Agenda

The agenda is rather simple. The project manager begins with an update on the status of the project. Any problems encountered are discussed and the solutions put in place are shared. If a particular problem does not admit of a solution, this is an opportunity for the project review team to suggest approaches. If there are any open issues from the prior project review, they are discussed and their status shared. All along, the review team may ask questions and otherwise engage the project manager in open discussion. While such an agenda may result in a review that is highly critical, it should also be supportive. Both parties need to approach a project review with that as an operating assumption.

4.1.3.3 Frequency

The frequency is irregular. Project reviews should occur at major milestones in the project. These can be selected from among all the project milestones to assure good oversight of the project and give the review team an opportunity to help the project team recover from significant problem situations. Project reviews can also occur at major funding points in the project; for example, after design has been approved and the project needs to get funding approval for the build phase.

4.1.3.4 Attendees

The review team should consist of four to six members. One should be from the management team in the PSO to assess compliance with standard practices and processes. One should represent and speak for the client or customer. The remaining members should be program and senior project managers. Their role is critical to successful project reviews. They can help the project manager not only with a review of the actions and decisions that were made but more importantly with problem solving. Having your peers take a look can be very beneficial and help everyone learn from the experience. Best practices can be discovered and shared.

4.1.3.5 Action Items

Based on the project status and the ensuing discussion there will most likely be a number of action items. These are usually directed at the project manager for resolution. It is expected that there will be a report on the action items at the next project review unless otherwise requested by the review team.

4.2 Prioritizing Improvement Opportunities

This section helps us build a prioritized list of improvement opportunities. While it applies primarily at the process-wide and practice-wide levels, it can also be applied to individual projects.

4.2.1 Ranking Improvement Opportunities

You will discover that there are more improvement opportunities than can be meaningfully worked on at any one time. What criteria should or could be used to decide which opportunities to pursue? Once the criterion is chosen, how should the opportunities be prioritized?

4.2.1.1 Forced Ranking Model

Of all the approaches this one is the simplest and perhaps the most subjective. As the number of improvement initiatives increases, so does the subjectivity of the process. That is a direct result of an overload of comparisons that the evaluator has to make. They simply cannot keep that many comparisons in mind at one time. As the number of reviewers increases, the objectivity of the process increases. The law of averages tends to smooth out any outliers from the rankings and the true rankings emerge. Each member of the evaluation panel is asked to rank all improvement initiatives based on their own criterion, or they may be required to use one that has been specified. The results are shown in columns 2 through 5 of Figure 4.6. The next step is to add up all of the ranks for a given initiative. That calculation is shown in the column titled "Rank sum." This column is used to determine the final ranking of all initiatives, which is displayed in the column titled "Forced rank." The lower the rank sum, the higher the

Improvement opportunity #	A	B	C	D	Rank sum	Forced rank
1	2	3	2	2	9	2
2	4	1	1	1	7	1
3	5	2	5	5	17	5
4	1	5	3	4	13	3
5	3	4	4	3	14	4

Figure 4.6 An example of a forced ranking.

priority of the improvement initiative. Ties are broken by choosing as the higher ranked initiative the one with the minimum poorest rank.

In the interest of equity, if there are multiple reviewers, they should agree on a criterion to use. However, a good case can be built for not specifying or requiring the use of the same criterion. The assumption underlying this statement is that each reviewer might use a different criterion and that any initiative that ranks high on a number of different criterion, is in fact a high priority initiative. The important factor is that whatever criterion the reviewer uses it must be applied consistently across all improvement initiatives they rank.

4.2.1.2 Risk Management Model

There are a variety of ways risk can be used to prioritize improvement initiatives. I prefer to use techniques that do not require extensive mathematical models to calculate risk. In my opinion, a risk probability of 0.682 and 0.713 are really the same and to try to prioritize based on such subjective metrics is a wasted effort, and you are really fooling yourself into thinking you have a repeatable metric.

Figure 4.7 is my preferred approach to prioritizing improvement opportunity using risk. There are five risk categories ranging from 1 (high probability of success) to 5 (low probability of success). High probability of success means low risk of failure. A rule must be established that categorizes the improvement opportunity into one of the five business categories and one of the five technology categories. The result is the two-way table shown in the figure. In the example data five improvement initiatives have been assessed on the two probabilities and plotted in the matrix. Those improvement initiatives that fall in the lightly shaded region (the northwest corner of the matrix) should be undertaken (#1 and #2). Those that fall in the darkest region should not be undertaken (#3 and #5). Those improvement initiatives that fall in the unshaded region (#4) would be undertaken as resources permit. The specific priority ranking of these five initiatives is 2, 1, 4, 3, and 5. The closer to the northwest corner, the higher the rank.

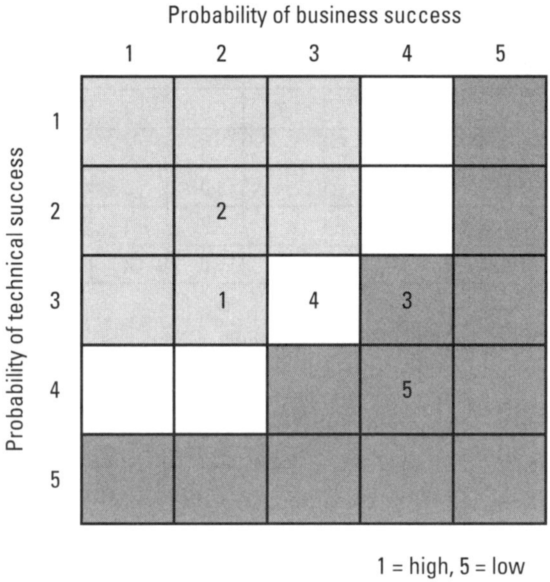

Figure 4.7 An example of the risk matrix.

The number of classes, five in our example, is variable. Three can be used, but often a smaller number does not provide enough discrimination between improvement initiatives. My choice is to use five.

4.2.1.3 Paired Comparisons Model

Figure 4.8 is an example of the use of the paired comparisons model. Every improvement initiative is represented by a single row and a single column. In this model every pair of improvement initiatives is compared. The one that is preferred by the reviewer is given a value of 1 and the other a value of 0. The "X" down the diagonal occurs because improvement initiatives are not compared against themselves. Let us use an example to show how the matrix is completed. We are going to compare improvement initiative #1 to all the other improvement initiatives. If #1 is preferred to #2, place a 1 in the (1, 2) cell and a 0 in the (2, 1) cell. Continuing in this fashion all of the off-diagonal cells will have either a 1 or a 0 placed in them. Note that if cell (i, j) is 1, then all (j, i) is 0. Sum each row and rank the improvement initiatives from the highest rank sum to the lowest rank sum. Ties are broken by examining the relationship between the tied improvement initiatives. In the example, #3 and #5 are tied for the fourth rank. Since #5 is preferred to #3, #5 is given rank of 4 and #3 is given the rank of 5.

The paired comparison model does not scale very well, but since there should be fewer than 10 potential improvement initiatives, it should not be too

Improvement opportunity #	1	2	3	4	5	Sum	Rank
1	X	1	1	0	1	3	2
2	0	X	0	1	1	2	3
3	0	1	X	0	0	1	5
4	1	1	1	X	1	4	1
5	0	0	1	0	X	1	4

Figure 4.8 An example of paired comparison model.

cumbersome. Ten improvement initiatives require 45 paired comparisons. The criterion that the reviewer uses is not specified. For consistency sake the reviewer should choose a single criterion and use it across all comparisons.

4.2.1.4 Weighted Criteria Model

This approach is the most robust of the quantitative approaches one might select. It allows the reviewer, or review board, to select the specific criteria and to weight the criteria by importance to one another. The example given in Figure 4.9 is quite general but it illustrates the point quite effectively.

Every improvement initiative is processed through the model and their scores are used to rank the initiatives—the higher the score, the higher the rank.

Criteria	Criteria weight	Very good (8)	Good (6)	Fair (4)	Poor (2)	Very poor (0)	Expected level weight	Expected weight score
Fit to unit mission	10	1.0					8.0	80.0
Fit to unit objectives	10	0.2	0.6	0.2			6.0	60.0
Fit to unit strategy	10			1.0			4.0	40.0
Contribute to PSO goal A	8				1.0		2.0	16.0
Contribute to PSO goal B	6	0.2	0.8				6.4	38.4
Contribute to PSO goal C	4		0.5	0.5			5.0	20.0
Takes advantage of strengths	10				0.6	0.4	1.2	12.0
Avoid weaknesses	10	0.7	0.3				7.4	74.0
Improvement opportunity #1								340.4

Figure 4.9 An example of the weighted criteria model.

4.3 Points to Remember

The following is a list of points to remember from this chapter:

- C/SC is a tool that can be used to dissect a project's performance with respect to actual versus planned schedule and budget.
- There are four basic ways to measure value:

  1. 100% allocated when the task is first open for work;
  2. 100% allocated when the task is complete;
  3. 50% allocated when the task is first open for work and the remaining 50% when the task is complete;
  4. Value as a percentage of the subtasks complete at the time of the report.

- The CPI is defined as the ration of EV to AC. Values above 1 indicate a below-budget situation.
- The SPI is the ration of EV to PV. Values above 1 indicate an ahead-of-schedule situation.
- The MTC allows one to track the trends in the scheduled date of a future event as project work commences.
- The MTC can be adapted to track trends in CPI and SPI as project work commences.
- Project reviews are held at periodic times (milestone events, for example) for the purpose of assessing the performance of the project and offering corrective action ideas where needed.
- The forced ranking model is a simple way to prioritize a number of improvement initiatives based on the individual rankings of a number of reviewers. It does not scale very well and becomes more subjective as the number of improvement initiatives increases. It becomes more objective as the number of reviewers increases.
- The risk management model is a two-dimensional way to prioritize a number of improvement initiatives based on categorical assessments of technology risk and business risk.
- The paired comparisons model requires the pair-wise comparison of each improvement initiative. It does not scale very well.
- The weighted criteria model is a quantitative model for scoring an improvement initiative on a number of weighted criteria. It is entirely flexible and adaptable to any situation, and it does scale very well.

Reference

[1] Wysocki, R. K., and R. McGary, *Effective Project Management, 3rd Edition,* New York: John Wiley & Sons, 2003.

5

Tools to Investigate Improvement Opportunities

We have discussed the assessment, plotting, and analysis of both PD and PP maturity level data. We know how to target those knowledge areas that should be further investigated for improvement opportunities. We have suggested a few approaches for ranking improvement opportunities. In this chapter we look under the hood at a knowledge area or individual process within a knowledge area to analyze the PD or PP performance to define potential areas for improvement initiatives. A number of commonly used quality assessment and problem solving tools are described.

5.1 Problem Solving for Continuous Improvement

Problem solving is fundamental to any serious investigation of improvement opportunities. There are several problem-solving models available. The one that I have used extensively and would like to focus on is a model I first introduced about 15 years ago [1]. I have used it for a number of years and feel particularly comfortable applying it to process improvement programs.

5.1.1 Definition

Figure 5.1 is a graphic portrayal of my problem-solving model. It was originally designed to support different situations that arise in project management and works equally well in process improvement programs. I have used it with great success for several years. This problem-solving model consists of the seven steps

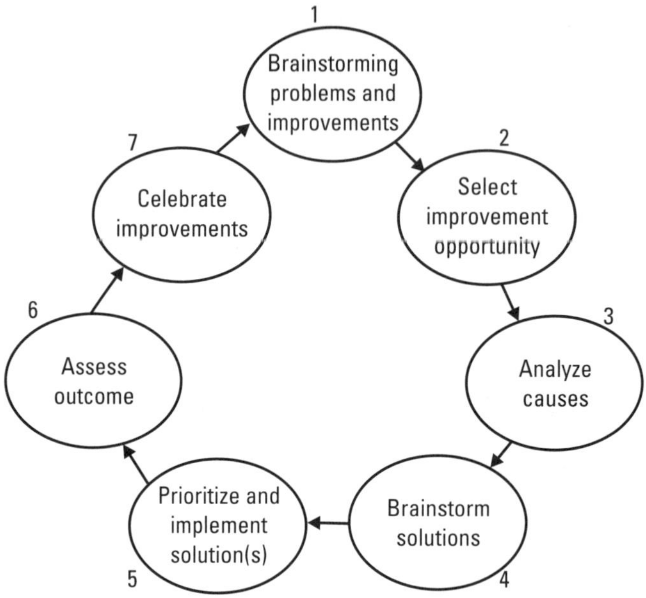

Figure 5.1 A problem-solving model for continuous improvement programs.

outlined in the figure. The model is quite robust and works regardless of the maturity level problems you are facing. The model works equally well at the knowledge area or for processes within a knowledge area level. Let us take a closer look at each step.

5.1.1.1 Step 1: Brainstorming Problems and Improvements

My assumption in this discussion is that you have formed a continuous improvement task force and it is that task force that is using the problem-solving model. Based on the task force members' specific understanding of the knowledge area or process within a knowledge area and its PD and PP maturity levels, the underlying problems and improvement opportunities might come to light through a brainstorming session. The benefit here is that the openness of a brainstorming session tends to bring out opinions and suggestions that might otherwise be held back. This is important because we do not want to miss any opportunities for improvement. Getting all of the task force's ideas on the table gives them the best chance of correcting the maturity anomaly.

5.1.1.2 Step 2: Select Improvement Opportunity

When the first part of the brainstorming session is complete, there should be a good number of ideas posted on the whiteboard. As the discussion of the posted ideas takes place, a consensus opinion of the most significant problem should emerge. That identifies the improvement opportunity that will be pursued. If more than one emerges, the task force can modify the remaining steps to account for more than one improvement opportunity and even more than one process can be targeted for improvement. For simplicity's sake we will proceed as though there were only one process and one improvement opportunity identified.

5.1.1.3 Step 3: Analyze Causes

With the improvement opportunity identified, the task force should begin to list the probable causes of the PD or PP anomaly. This could be done with a brainstorming session, but that has not proved to be any more productive or efficient than a simple discussion of probable causes. I will discuss some tools that I have used for this step later in the chapter.

5.1.1.4 Step 4: Brainstorm Solutions

Either a brainstorming session or a simple discussion of possible solutions can be done here. The choice is a matter of what works best for the situation at hand. In any case a small number of possible solutions to the anomaly will be listed.

5.1.1.5 Step 5: Prioritize and Implement Solution(s)

Any of the prioritization tools introduced in Chapter 4 can be used to rank the solutions. Every attempt should be made to produce an untied ranking so that a single solution can be implemented, its outcome measured, and additional solutions implemented as needed. This keeps life simpler for all involved especially for the project teams that have to implement any changes that are forthcoming from the initiative. If it should happen that the two top ranked solutions are independent of one another, then both might be pursued concurrently. Implementing a solution is a project. It is usually a short-term project. These improvement initiative projects are quite volatile and subject to change and cancellation. That is their nature. Think of these projects as pursuing the best suggested solution and subject to learning and discovery as the project work commences. A new or revised solution may suggest itself. That calls for changes in the project plan. In extreme cases it may call for a cancellation of the project and the commissioning of a new improvement initiative following a new and completely different solution.

5.1.1.6 Step 6: Assess Outcomes

As part of the planning for implementation of the solution(s), measures of success must be established. The solution may be such that the entire maturity anomaly may be removed or only part of the anomaly removed. The project team that undertakes the solution must clearly know what is expected of the improvement initiative.

As the project work commences, the progress towards the success criteria should be monitored. Poor progress may lead to cancellation of the solution because it just is not producing the expected results. The solution may not be appropriate for the problem at hand and another approach must be undertaken.

5.1.1.7 Step 7: Celebrate Improvements

Any improvement that meets the success criteria should be celebrated. This does not mean having an elegant dinner party at a local five-star establishment. What it does mean is that there is some recognition of what was done and who did it. Do not forget to recognize any project teams that suffered through the changes with the task force. Implementing change in an organization is hard work and when it is successful, all parties that participated in it deserve the appropriate notoriety. This closes the improvement loop and the task force returns to Step 1. The same process may still be the subject of their attention, and that depends on whether the overall improvement goal has been reached. It is also possible that while pursuing one improvement initiative another one is discovered. That will lead to a follow-up initiative.

5.2 Brainstorming

We are all familiar with brainstorming and so this will be a short description just to make sure we are all on the same page when it comes to running a brainstorming session. A session begins with the facilitator stating the reason for the session. Often it will be a problem that needs a solution. The facilitator invites the attendees to offer any thoughts, ideas, or suggestions that may be relevant to the solution. It is important that the attendees realize that all contributions are welcome and that there will be no criticisms allowed. Contributions are merely recorded for all to see. The only discussion that is allowed is for clarification of some point that was raised. I have often seen one idea lead to another and another and before your realize it you are converging on a solution. When no new ideas are forthcoming the session is ended. It is often followed by a grouping of like comments, a prioritization of the suggestions, and perhaps a formulation of next steps.

5.3 Fishbone Diagrams

The fishbone diagram, also known as cause and effect diagrams and Ishikawa Charts, is a graphical representation of the possible causes of an outcome or effect.

Figure 5.2 is a generic version of the fishbone diagram. The box at the right end describes the effect or outcome that is being targeted. The ellipses identify the major categories within which the causes are to be defined. Along the line leading from a major cause category to the horizontal arrow are listed all of those specific causes within the cause category that might be contributing to the effect.

5.4 Force Field Analysis

One step up from fishbone diagrams is force field analysis. On balance it should contain all of the causes that were identified through the fishbone diagram, which are called restraining forces in force field analysis. Restraining forces are forces that act upon the effect to maintain it or prevent an unacceptable situation from improving. An additional set of positive factors, which are called driving forces, have the opposite impact of restraining forces. These forces act upon the effect to neutralize it or to improve the situation. The goal of a force field analysis is to identify these two types of forces and then identify initiatives to reduce the restraining forces and strengthen the driving forces. Together the restraining forces and driving forces are the causes that act upon each other to produce the current state—the effect. Figure 5.3 is a generic graphical representation of this model.

Once these forces have been identified, the next step is to identify actions that might be taken to weaken or remove one or more of the restraining forces

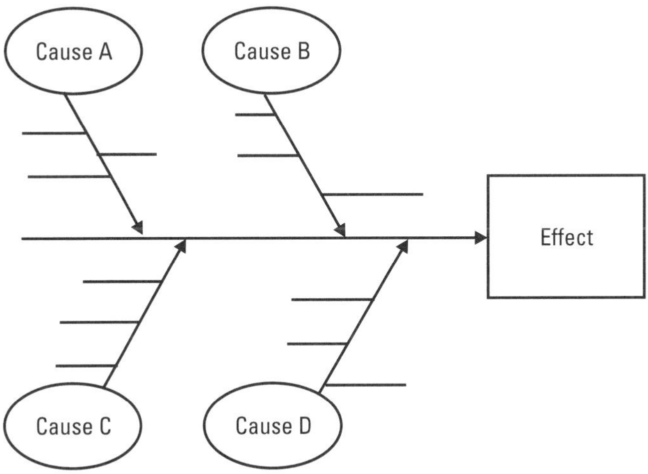

Figure 5.2 The structure of the fishbone diagram.

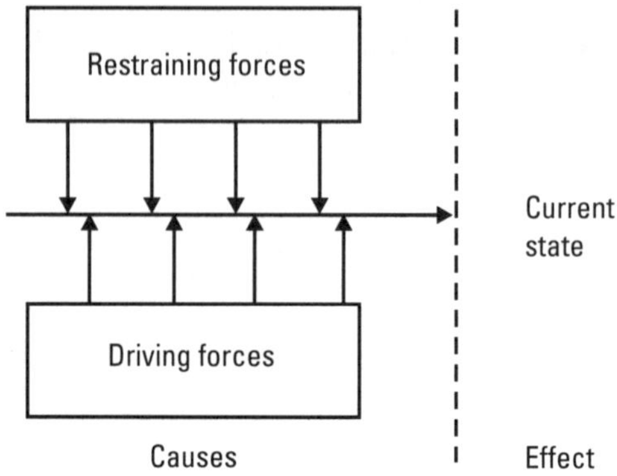

Figure 5.3 Force field analysis.

and strengthen or complement one or more of the driving forces. These actions become a set of initiatives that will be rank ordered and undertaken sequentially.

5.5 Pareto Diagrams

The Pareto diagram is a simple yet elegant approach to problem resolution. It identifies and prioritizes the causes of the problem by their frequency of occurrence and attacks them in that order.

Pareto analysis begins with a survey and collection of data focusing on the causes of some unacceptable outcome. A Pareto diagram can be used quite effectively to help identify areas where PP improvement initiatives may be productively undertaken. Figure 5.4 is a generic representation of the results of the

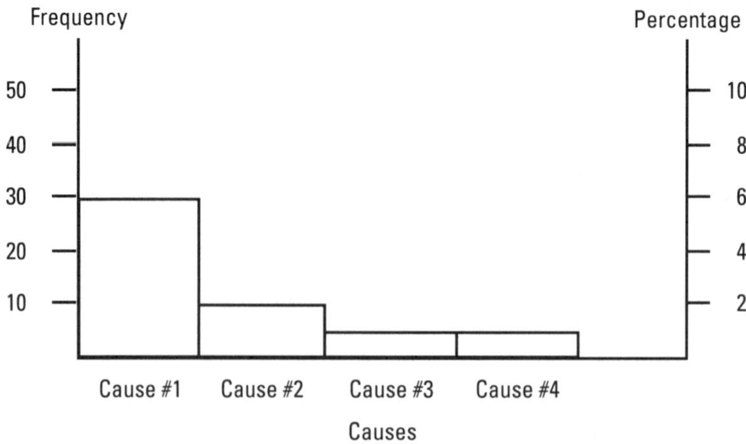

Figure 5.4 Pareto diagram.

collection of data on relative frequencies of causes of the problem under consideration. The causes may be defined ahead of time but are generally identified as part of the data collection activity.

5.6 Process Charts

There will be situations where the object of an improvement initiative may be an entire process. For those situations you will need a graphical method of describing the process. I prefer graphical approaches that are simple and intuitive and have found the process charts to be exactly that.

A process chart is a graphical picture of a process that is comprised of a number of steps. These steps may be decision points, data collection points, approvals, or report generation. Each of these types of steps is represented by a different icon. The icon stylistically portrays the activity associated with that step, which adds to the intuitiveness of the process chart. Figure 5.5 is an example of a process chart that utilizes the more commonly used icons. It shows a typical order entry process. Note that it has been arranged by customer, sales, credit, and fulfillment. Each of these is called a swim lane and it extends horizontally across the chart. The major units or individuals that interact with the process are each represented by a swim lane. Swim lanes are stacked vertically in

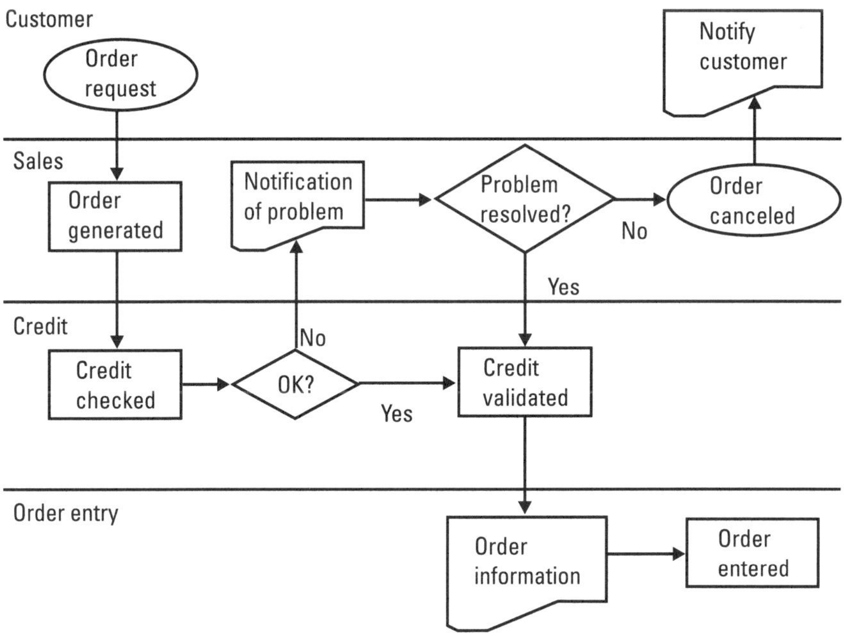

Figure 5.5 A typical process chart for an order entry process.

more or less chronological order. If you move horizontally across a swim lane, you will see all of the actions of that swim lane within the process. Standard flow charting symbols are used in the process chart. The oval represents the beginning and ending activities of the process. The rectangle is a process step. The diamond is a decision point. The page icon is a report. The flow in a process chart is top to bottom and left to right wherever possible.

5.7 Root Cause Analysis

Knowing that an undesirable situation exists is the starting point for root cause analysis. Root cause analysis is represented as a hierarchical chart and is very intuitive. Starting from the top, let us define the structure and terminology of root cause analysis. At the top is the problem statement. The only restriction is that it must be a single problem. It must be clearly stated. Avoid jargon or terminology that might not be understood by anyone who might have the occasion to read it. The second level consists of one or more reasons for the problem. You must have the assurance that the reasons you state really are the reasons and not someone's conjecture. The reasons should be commonly accepted by the organization so that they do not need to be defended. The next several levels are "why" questions. There may be more than one why question per reason and more than one answer per why question. The answer to one why question gives rise to one or more why questions. At some point in this hierarchy an answer becomes a statement for which no further question makes sense. You might call this a cause. Figure 5.6 is a generic representation of a root cause analysis.

5.8 Prioritizing Processes

In order to implement improvement initiatives, you must first have some way of prioritizing them. That will allow you to tackle them in an organized fashion. You may choose to schedule improvement initiatives by knowledge area, in groups, or one at a time.

5.8.1 Scheduling Improvement Initiatives by Knowledge Area

The first pass at that prioritization by groups was discussed in Chapter 3 and presented in the zone map in Figure 3.12. Recall that the zone map was generated from the (PD or PP) maturity level of a process and its correlation factor. As simple as those metrics are, they allow us to crudely prioritize the 39 processes. Basically, the closer a process is to the northwest corner of the zone map, the higher the priority of the process as a candidate for improvement.

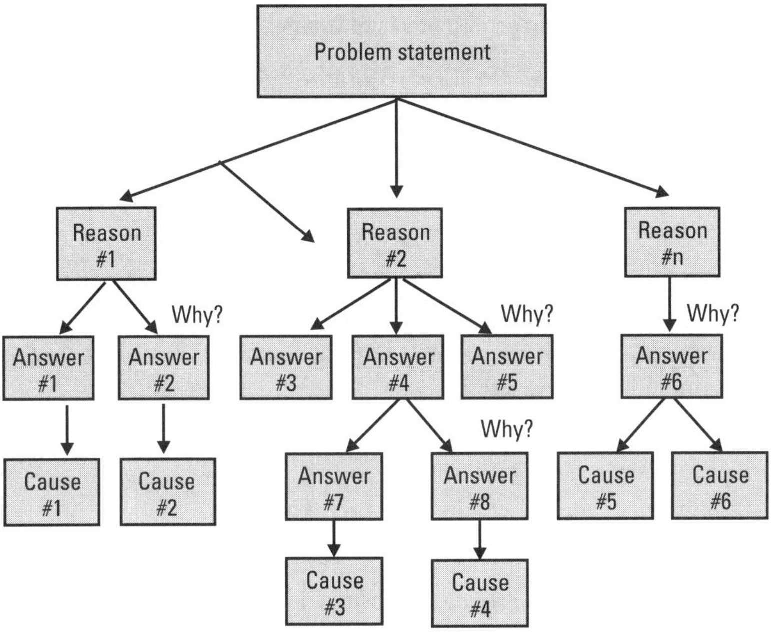

Figure 5.6 Generic representation of a root cause analysis.

At the program level you will want to identify the knowledge area most in need of improvement. That will be the knowledge area whose processes are nearest the northwest corner of the zone map. In addition to its location on the zone map, that need may be based on project performance, business needs, or any other criteria that can be used to rank order the knowledge areas.

By inspecting Figure 3.12 you will note that HR management (P21, P22, and P23) has the lowest PD values of all the processes (1.14, 1.22, and 1.10, respectively). Cost management (P14 = 1.50, P15 = 1.38, P16 = 2.00, and P17 = 1.22) and risk management (P28 = 1.33, P29 = 1.00, P30 = 2.00, P31 = 2.00, P32 = 1.33, and P33 = 2.00) are not far behind. That means that your improvement programs should focus on HR, cost, and risk in that order and preferably one at a time.

If you wish to focus your improvement programs on PP values, then Figure 5.7 would be changed to display the PP data for your prioritization decisions. Figure 5.7 is a plot of the average PP maturity level for nine projects that recently failed. Here you will note that all five of the scope management processes (P04, P05, P06, P07, and P08) have maturity values in the 2+ range and correlation factors of 6 and above. This would clearly be your first choice for improvement initiatives aimed at increasing the PP values. The second choice should be HR management (P21, P22, and P23). All of those PP values are in the 1+ range and have correlation factors of 5 and above.

5.8.2 Scheduling Improvement Initiatives in Groups

The reason you would want to form a group of processes for an improvement initiative is that they possess the following:

- They are grouped in the zone map in the same or neighboring cells.
- They form a dependent set of processes.

Any change to one process will most certainly impact any related processes. That relationship may be a causal one or one in which the output of one process becomes the input to one or more other processes. Furthermore, a change in one process may require some form of complementary or coordinating change in another process.

5.8.3 Scheduling Improvement Initiatives One at a Time

Both Figure 3.12 and Figure 5.7 can be used as the basis for calculating a metric that will allow you to create a single prioritized list of the 39 processes. Since a

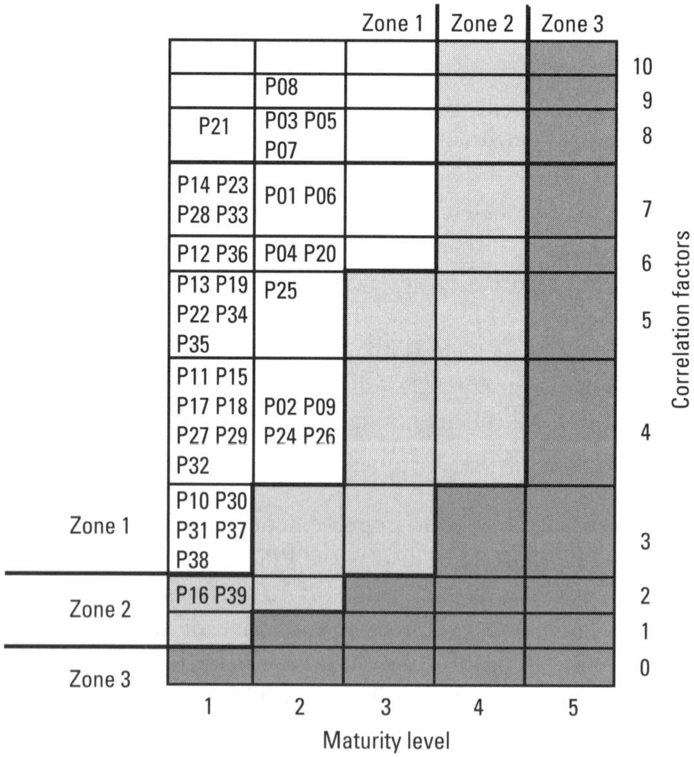

Figure 5.7 The PP maturity zone map.

prioritization of the 39 processes is not immediately suggested by either Figure 3.12 or 5.7, another filter is needed. In this case the product of the calculated maturity level of the process and (10 minus its correlation factor) produces a natural ordering of the 39 processes. If one or more processes were correlated with all 10 CSFs, then use 11 instead of 10 in the formula. Applying the formula results in the prioritized list for the 39 processes shown in Table 5.1.

There are at least two ways that this prioritized list can be used. The first would be to establish target maturity levels for each process and then begin your improvement project with the first process on the list. The improvement project goal is to raise the PD (or PP) maturity level value to the target value. The second approach would be to implement a single improvement project for the first process on the list, complete the project, assess its new PD or PP value, recalculate the formula value of the now improved process and return it to the list in its new prioritized position. The next improvement project would be with the now first process on the revised list. This approach can be repeated until all processes reach their target maturity level. In actual practice your improvement program would most likely involve a subset of the nine knowledge areas. An example is given in Chapter 7.

Table 5.1
Priortized List of Improvement Areas

Process Number	Process Name	Knowledge Area
P21	Organizational planning	HR
P23	Team development	HR
P08	Scope change control	Scope
P28	Risk management planning	Risk
P14	Resource planning	Cost
P19	Quality assurance	Quality
P03	Integrated change control	Integration
P06	Scope definition	Scope
P07	Scope verification	Scope
P18	Quality planning	Quality
P29	Risk identification	Risk
P33	Risk monitoring and control	Risk
P22	Staff acquisition	HR

Table 5.1 (continued)

Process Number	Process Name	Knowledge Area
P17	Cost control	Cost
P32	Risk response planning	Risk
P05	Scope planning	Scope
P36	Solicitation	Procurement
P12	Schedule development	Time
P15	Cost estimating	Cost
P01	Project plan development	Integration
P10	Activity sequencing	Time
P13	Schedule control	Time
P20	Quality control	Quality
P35	Solicitation planning	Procurement
P11	Activity duration estimating	Time
P24	Communications planning	Communications
P26	Performance reporting	Communications
P30	Qualitative risk analysis	Risk
P31	Quantitative risk analysis	Risk
P37	Source selection	Procurement
P38	Contract administration	Procurement
P25	Information distribution	Communications
P34	Procurement planning	Procurement
P04	Initiation	Scope
P16	Cost budgeting	Cost
P39	Contract close-out	Procurement
P09	Activity definition	Scope
P02	Project plan execution	Integration
P27	Administrative closure	Communications

5.9 Recap

By way of a reference guide, Table 5.2 shows which tools apply to which steps in the problem-solving model. The tools from Chapters 4 and 5 that are shown in the table are listed in the key below.

5.10 Points to Remember

The following is a list of points to remember from this chapter:

- The problem-solving model is a seven-step model:
    1. Brainstorm problems and improvements;
    2. Select improvement opportunities;
    3. Analyze causes;
    4. Brainstorm solutions;
    5. Prioritize and implement solution(s);
    6. Assess outcomes;
    7. Celebrate improvements.

Table 5.2
Use of Tools For Each Step

Step	Tools
Step #1: Brainstorming problems and opportunities	B
Step #2: Select improvement opportunities	FRM, PCM, RMM, WCM
Step #3: Analyze causes	FD, FFA, PA, PC, PR, RCA
Step #4: Brainstorm solutions	B
Step #5: Prioritize and implement solutions	FRM, PCM, RMM, WCM
Step #6: Assess outcome	CSC, MTC, PR
Step #7: Celebrate improvements	On your own

B: brainstorming; CSC: cost/schedule control; FD: fishbone diagram; FFA: force field analysis; FRM: forced ranking model; MTC: milestone trend charts; PA: Pareto analysis; PC: process chart; PCM: paired comparison model; PR: project reviews; RCA: root cause analysis; RMM: risk management model; WCM: weighted criteria model.

- The fishbone diagram (a.k.a. cause and effect diagram, Ishikawa Chart) is a graphical tool for identifying the probable causes of a situation.
- Force field analysis is a graphical tool for explaining the restraining forces that tend to sustain a situation and the driving forces that tend to improve the situation.
- Pareto diagrams graphically display the percent attributable to each of several causes of a situation and thereby highlight those causes that are more prominent and candidates for further investigation.
- Process charts are graphical pictures of a process made up of several interdependent steps and are used as input to a process design change.
- Root cause analysis is a tool that repeatedly asks the question "why" in order to discover the true cause of a situation.
- There are basically three ways to prioritize processes for improvement initiatives:

    1. By knowledge area;
    2. By groups of dependent processes;
    3. By individual processes.

Reference

[1] Wysocki, R. K., and R. McGary, *Effective Project Management, 3rd Edition,* New York: John Wiley & Sons, 2003.

6

Commissioning Improvement Initiatives

At this point we have assembled all of the tools we will need to implement a continuous project management process improvement program. Our next task is to put all of this together into a coherent program that moves our project management culture from its current maturity level to a desired end state. That end state may encompass all 39 project management processes, all nine knowledge areas, or only a selected number of processes or knowledge areas. The choice is more a matter of responding to the business situation that has been created by a higher than expected project failure rate than anything else.

6.1 Characteristics of an Improvement Program

An improvement program is a collection of improvement initiatives that are related to one another because they are all focused on the processes within a single knowledge area. Multiple improvement programs, each focusing on a different knowledge area, may be conducted concurrently. This multiple program effort is undertaken to raise the maturity level of the entire project management environment. Such programs are high risk and should be undertaken only in dire circumstances. The goal of a single improvement program is to raise the maturity level of a single knowledge area to some specific level. That will happen through a number of projects that are all dependent upon one another and by focusing on different processes within the knowledge area. In order for the knowledge area to reach a certain maturity level, all of the processes within that knowledge area must be at or above the targeted maturity level.

6.1.1 Long Duration

Improvement programs may be budgeted for a specific time and cost, but to reach the goal level of maturity the actual program may be longer or shorter than planned. Improvement programs always involve change, and it is hard to timebox how long it will take for the change to be integrated into the organization and affect project performance and success rates. In the case of programs, the goal, which is one of the triple constraints, is fixed, and that means that either or both time and cost, which are the other two constraints, must be variable. With some exception the cost of an improvement program is mostly tied up in labor. While there may be an opportunity loss associated with tying up that labor, many improvement programs can use professionals that are not on billable projects. In this case their time is a sunk cost, and so allocating them to improvement programs will not carry any real costs.

6.1.2 Multiproject Approach

Improvement programs are really projects of projects. Because many of these projects will be dependent upon one another, an improvement program can be represented as a precedence diagram. Figure 6.1 is an example. There is a difference however. Some improvement projects may not be executed. Their predecessor projects may have reached the goal for the processes they are trying to improve, and therefore, succeeding projects for that same process will not be necessary. Resources can be diverted to other improvement projects in the

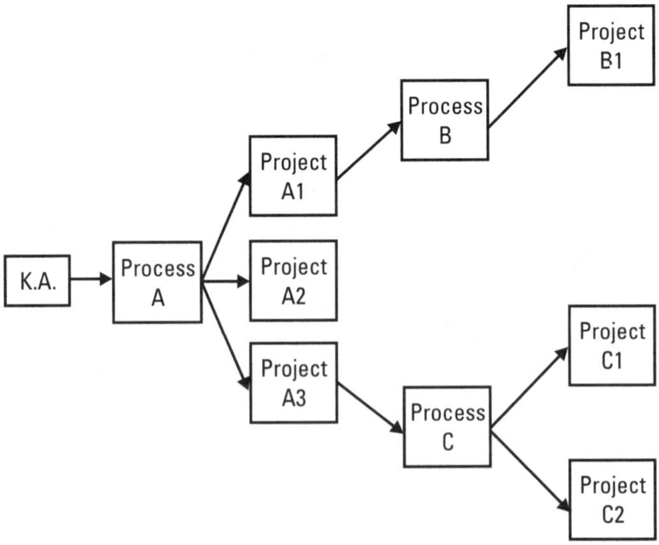

Figure 6.1 A generic program project precedence diagram.

program. Similarly, the completion of one project may suggest a change in one or more later projects or the addition of one or more projects to the program.

Figure 6.1 is the complete improvement program for the chosen knowledge area. All six projects are defined in terms of expected results but are not yet planned. Their planning is contingent on the results of Project A1 and A3.The first three improvement initiatives are related to Process A. They are Project A1, A2, and A3. Depending on the results from Project A1, the Process B project, Project B1, may or may not be launched, and if it is launched, its expected result may be different than originally planned, and exactly what will be done in Project B1 then defined and planned. The same is true for Process C. Depending on the results from Project A3, Project C1 and C2 may or may not be launched and may have revised expectations. While Figure 6.1 looks like a precedence diagram, it differs in that many of its projects are done on a contingency basis or not at all. For example, the results of Projects A1, A2, and A3 may be such that further improvement projects are not advised.

6.1.3 Just-in-Time Planning

Projects within an improvement program are not planned until it is decided that they will actually be executed. Why waste time planning a project that might be canceled before it is even begun?

6.1.4 High Change

At the improvement program level, change occurs as learning and discovery from early projects suggests that later projects be removed from the program or be modified. In many cases new projects are identified and added to the portfolio of projects in the program. Remember that the goal of an improvement program is focused on raising the maturity level of a specific knowledge area. That means that several processes are the target of an improvement program. Each process will have multiple projects associated with it, and the success or failure of one project can have a ripple effect through all of the projects that pertain to that process. In other words, Figure 6.1 is a dynamic structure that changes continuously as individual improvement initiatives are undertaken or completed. That suggests a just-in-time planning strategy.

6.1.5 High Kill Rate

There is nothing wrong with killing projects in an improvement program. Resources should be used in the most effective and efficient manner possible. The final goal of the improvement program should always be kept in mind as individual projects are assessed as to what and how they can contribute to that goal. The management of an improvement program is not any different than the management of a project portfolio. In the general case the objective of the portfolio is to maximize return on the portfolio. In the case of an improvement program the return on investment is measured in terms of maturity level improvements. The mix of projects in an improvement program is chosen so as to maximize the impact on the maturity level of the knowledge area defined for the program. If project performance indicates that an improvement objective is not likely to be met, kill the project and assign the resources to more promising initiatives.

6.2 Characteristics of an Improvement Initiative

An improvement initiative is a project within an improvement program that is designed to impact the maturity of one of the 39 processes that define a project management culture.

6.2.1 Short Duration

While an improvement program may be a very complex and lengthy effort, a single improvement project is short-term. It has a very narrowly defined scope. In and of itself it will not make a significant impact on the maturity level of the process(es) on which it is focused. However, in the larger context of the program it is just one piece of the puzzle. Every improvement initiative has an expected success criteria, which is usually expressed in terms of an impact on the maturity level. The performance of the improvement program compared to its expected result is the criteria for killing or continuing the project.

6.2.2 Multiphase Approach

So far we have been discussing the most primitive of improvement initiatives—the individual project. The next level of simplicity will be a series of phases within the project each focused on improving the same process but from a different perspective. In other words a single improvement project can include different approaches to the same problem. These approaches are most likely dependent upon one another, as illustrated in Figure 6.2

When it is decided that an improvement program for Process A is to be undertaken, a prioritization of 12 improvement initiatives is established. They are dependent upon one another as shown in Figure 6.2. The same reasoning that led to the contingencies in Figure 6.1 applies here. For example, Projects A1, A2, and A3 are run concurrently. The results of A1 and A2 determine if and how Projects A1.1 and A2.1 will be conducted. The results of Project A2.1 will determine which of Projects A2.1.1 and A2.1.2 will be done and how. As in Figure 6.1, all project planning is done just-in-time.

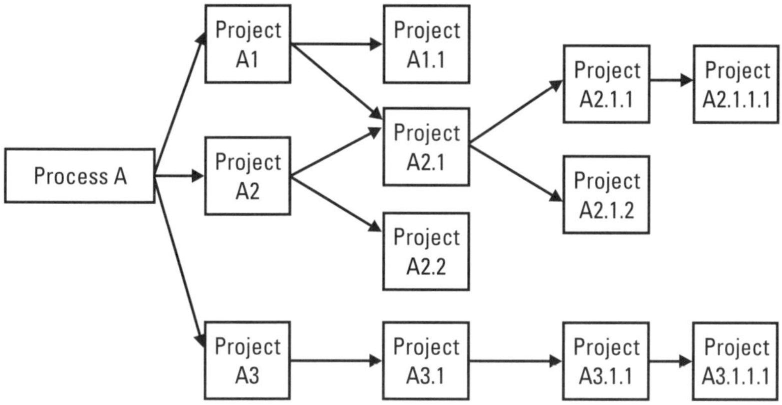

Figure 6.2 A series of dependent improvement initiatives.

6.2.3 Just-in-Time Planning

Improvement initiatives at the project and phase level are volatile. By that I mean they are frequently canceled because they are not producing the improvement results expected, or the learning and discovery that takes place during the project work render the current project plan obsolete. That means change and plenty of it. The best way to accommodate that change is to only plan parts of the project and see where that leads you. You may discover that the original direction is bearing fruit and should be continued—in which case extend the plan. Or you may discover that a different direction looks more promising—in which case extend the plan in that direction. The message I want to send is to not spend a lot of time planning a project that is very likely to change or be canceled in midstream.

6.2.4 High Change

Just as discussed above, improvement initiative projects are subject to frequent change. That change will occur not only between projects that are dependent on one another but also within a project.

A sequence of projects may be established that all point at the same improvement initiative. The projects will usually be ordered from estimated greatest improvement to estimated least improvement. Each project is expected to contribute its part to the improvement. Collectively all of the projects in the sequence are expected to meet an overall improvement goal. Once a project is complete there will be an assessment of what improvement actually occurred compared to what improvement was expected. Four decisions can come from this analysis:

1. The overall improvement goal has been met and the sequence can be discontinued.
2. Everything is performing according to plan and the sequence should be continued.
3. The improvements are far below expectations and the sequence should be canceled.
4. The actual improvement of the most recently completed project leads us to revise the remaining projects in the sequence.

A single project is underway and some significant learning and discovery has taken place. There are two possible decisions that might be taken:

1. The project may not be yielding the improvement expected and should be canceled.
2. The project may be yielding a much greater improvement than expected, which leads the team to consider other improvement opportunities that had not yet been considered.

6.2.5 High Kill Rate

I think it is best to view improvement initiatives as exploratory. You certainly spent some time with your colleagues brainstorming and prioritizing improvement alternatives. Collectively your group made a decision as to which initiative(s) to pursue. You made that decision based on a hunch that the effort would be rewarded and the maturity level of a process or knowledge area would be positively impacted. Maybe it will, maybe it won't. Because these initiatives are exploratory, you can expect a high failure rate. Many of them will have to be abandoned because a better approach was discovered as part of the work of the initiative. Do not view that as a sign of failure, it is not. It is a sign that learning is taking place and that you are converging on an initiative that will have payoff.

6.3 Setting Maturity Goals

Not every project management process needs to be at level 5 maturity. Maybe none of them needs to reach level 5. Also, not every project management process needs to reach the same level of maturity. Setting the maturity goals for your project management process is more involved than you might have originally thought. Let us take a brief look at some of the factors that need to be considered.

While you certainly want to have the best project management processes and practices you can, the cost or attaining and maintaining them may not be practical. Given that level 5 maturity is out of reach for all 39 processes, what compromise position makes sense? I cannot envision any organization that is serious about cultivating a project management culture that would settle for less than Maturity Level 3 for all 39 processes. That means whatever your PD maturity levels might be, your PP maturity level values are all at or above 3.0. Period. That goal is a difficult one to achieve. Stop and think about the infrastructure you will need to certify that you are operating at level 3 maturity. Every project must undergo a series of project reviews at key points and milestones along the project life cycle. Any anomalies must have a corrective action plan developed, put in place, and monitored for compliance. Process documentation must be complete, clear, readily available, and backed up with the needed training and consulting support. And then you must staff for delivery of all these support services.

6.4 Scope the Initiative

The initiative starts out as an idea that must be documented. A tool that I developed and have used for many years as a scoping document in initiating project management projects is called the project overview statement (POS). This section describes each of the five items that make up the POS.

6.4.1 Evaluating Improvement Opportunities

Since there may be several improvement opportunities that are suggested by the data, it is important that we have an equitable way of comparing them. We will adapt the POS [1]. A full detailed discussion of the POS can be found in the cited reference there. In this section we present the basic ideas behind the POS so that it can be used without the need for further details.

The POS is a one-page description of the improvement opportunity. It has five parts as described below. Figure 6.3 is the POS template.

6.4.1.1 Improvement Opportunity Statement

This section defines a specific improvement opportunity that is supported by PD or PP maturity data, or by a recently completed individual project review. It is important that this section of the POS be grounded in established fact so that no one can legitimately challenge the improvement opportunity statement. A brief capsule description of the current situation may be described as well. The corporate expectations for the identified process might also be included to place some perspective on the value or importance of the opportunity to the

PROJECT OVERVIEW STATEMENT	Project Name	Project No.	Project Manager
Problem/Opportunity			
Goal			
Objectives			
Success Criteria			
Assumptions, Risks, Obstacles			

Prepared by	Date	Approved by	Date

Figure 6.3 POS template.

organization. Note that the improvement opportunity is stated using established data points for reference. This is a statement that is grounded in fact and not likely to be challenged by anyone in the organization.

6.4.1.2 Goal Statement

This is a brief statement of the goal that has been set for this improvement initiative. In the example the goal is stated clearly. It is unlikely that anyone can misinterpret the statement. Furthermore, at the end of the project there will be no doubt that the goal has or has not been reached. It is the type of statement that is clearly a "yes, we did it" or "no we didn't do it." It is critically important that there is a clear understanding and that there is no room for private interpretation. We do not want any eleventh-hour debates.

6.4.1.3 Objective Statement

The goal is further clarified in this section. Both the goal statement and the objective statement clearly bound the improvement initiative. In the example, the objective statements give a high level chronology of what the project will look like. The high level WBS can be started using the objective statements as the first level breakdown.

6.4.1.4 Success Criteria

This should be a quantitative statement of the outcome that is expected to result from this improvement initiative. It may simply be stated as the impact it will have on the maturity level of the affected process and the timeframe within which this improvement can be expected. Just like the goal statement, the success criteria are measurable outcomes. It either happened or it did not, and there is no room for private interpretations or misunderstanding. The success criteria is closely linked to the goal statement in the following way. The goal statement may contain a statement of success, in which case the success criteria would expand on the goal statement. It is important that the success criteria include measurable PD and PP results that will accrue as a result of the improvement initiative. These will often be the criterion used to prioritize improvement initiatives.

6.4.1.5 Assumptions, Risks, and Obstacles

This could be several statements of the basis on which this improvement initiative is undertaken and the potential barriers to success. In many cases the approval authority may be able to mitigate some of the risks and obstacles. These are fairly obvious and should appear in every POS related to PP improvement initiatives.

6.5 High-Level Planning of the Initiative

Because an improvement initiative may not be much more than a hunch, we do not want to spend anymore time planning than is necessary. Just-in-time planning, which is not an oxymoron, is an approach that will avoid needless expense of time and dollars. In this section we take a look at how that might be implemented within the context of an improvement project or an improvement program.

6.5.1 Work Breakdown Structure

Why would you want to waste time planning the future when the future is unknown? That same reasoning applies to creating the WBS for an improvement program. It even applies to an improvement project. My advice is to create a high-level WBS for the improvement program and a midlevel WBS for the highest priority project in the program.

6.5.2 Prioritize and Schedule Approaches

Using the high-level WBS we can prioritize all of the suggested improvement initiatives. As I discussed earlier, the prioritization should be based on which initiatives show the greatest promise of returning improvements. In some cases, the team may wish to pursue more than one initiative concurrently. This is acceptable as long as the initiatives are independent of one another. That is not a necessary condition, but it does make life simpler for the team or multiple teams if each team is working on a different initiative. The downside risk is that the initiatives will result in too many process changes to be integrated by the project team at one time. A more deliberate change strategy might reap better results.

6.6 Monitoring the Initiative

Each initiative must have a quantitative success criteria and it must be expressed in terms of maturity level impact. There may be other success criteria, but maturity level impact has to be one of them.

6.6.1 Define Performance Metrics

We are only going to concentrate on impacting the maturity level of a process or knowledge area. As stated above, there may be other success criteria but they are outside the scope of this book. At the program level the performance metric may apply to either a knowledge area or to a process within a knowledge area. At the project level the performance metric should refer to a process within a knowledge area.

6.6.1.1 Knowledge Area Performance Metric

Program improvement initiatives typically focus on a single knowledge area and have as their goal to raise the PP maturity level to a targeted value. Because the knowledge area will contain several processes and the project improvement initiatives will focus on individual processes; these programs will be long term. There will most likely be parallel initiatives throughout the program.

6.6.1.2 Process-Level Performance Metrics

Project improvement initiatives typically focus on a single process within a knowledge area and have as their goal to raise the PP and/or PD maturity levels to a targeted value. These initiatives will be short term as compared to the program improvement initiatives. In most cases there will not be parallel initiatives running concurrently because of the dependencies that will exist among improvement ideas.

6.6.2 Track Performance Metric

Once you have decided on what metrics you will track, the question becomes: how frequently? Too frequently and you may be micromanaging and adding too much nonvalue added work to the project team. The team may perceive your zeal as nothing more than micromanaging. Too infrequently and you may miss an opportunity to correct a situation before it gets seriously out of hand. Somewhere in between these two alternatives lies the best choice. My preference is for a weekly report. At the end of each work week I would ask for an update of the project file by all team members who were responsible for tasks that were open for work that week. If they update the project file in the midafternoon of the last work day of the week with where they expect to be at the end of the workday that leaves the project manager with the time to generate the performance metrics and report them by the beginning of the next workday. If you are holding 30-minute daily status meetings, you can report the performance metrics at the team meeting the day after.

6.7 Redirecting the Initiative

Improvement initiatives are the team's best guess at what will improve the PD or PP maturity level. As we discussed earlier, there will be a period of learning and discovery associated with these initiatives. Expect the learning and discovery to point out better approaches, especially for PP improvement initiatives. There are two general outcomes that you can expect: abandonment of an initiative or reprioritization and rescheduling of several initiatives.

6.7.1 Abandonment of Approaches

You may be laboring under the myth that to kill a project is a sign of failure. Nothing could be farther from the truth when it comes to improvement initiatives. I see them as a sign that learning and discovery has taken place and that resources are being redirected towards more fruitful efforts.

6.7.2 Reprioritize and Reschedule Approaches

Change is the mark of an improvement program. As we learn and discover what works and what does not work, we should reconsider the remaining initiatives in the program. Do some of them hold a higher promise of maturity improvement and should they have a higher priority among the remaining initiatives? Do some of them hold a lower promise of maturity improvement and should they have a lower priority or no priority among the remaining initiatives?

6.8 Closing the Initiative

The project work is complete and the project closing activities are begun.

6.8.1 Assess Final Performance Improvement

The changes from all of the completed improvement initiatives are documented and the project teams begin to implement them. This is a culture change that may be simple or may be dramatic. But at some future date you will collect PP maturity data and discover that you may or may not have reached the improvement level specified in the success criteria. Depending on the priority of the knowledge area or process, that may trigger another round of improvement programs and projects in an attempt to close the gaps.

6.8.2 Reprioritize Improvement Opportunities

With the outcome of the completed initiative now known and all learning and discovery from that project documented as part of the closing activities, it is time to reconsider any remaining improvement ideas in the program. The task force will have learned a great deal about how their initiatives were carried out and received by the teams. Maturity level targets will probably be revisited. A different rank order of knowledge areas and/or processes will be in place. Business conditions will have changed. All of these factors are input to help senior managers decide on the improvement programs and projects to be undertaken.

6.9 Points to Remember

The following is a list of important points to remember from this chapter:

- An improvement program is a collection of improvement initiatives that are related to one another because they are all focused on the processes within a single knowledge area.
- Improvement programs always involve change and it is hard to timebox how long it will take for the change to be integrated into the organization and affect project performance and success rates.
- Projects within an improvement program are not planned until it is decided that they will actually be executed. Why waste time planning a project that might be canceled before it is even begun?
- The management of an improvement program is not any different than the management of a project portfolio.
- Not every project management process needs to be at level 5 maturity.
- The POS is used to scope out the improvement program or project. It has five parts:

    1. Problem/opportunity statement;
    2. Goal;
    3. Objectives;
    4. Success criteria;
    5. Assumptions, risks, obstacles.

Reference

[1] Wysocki, R. K., and R. McGary, *Effective Project Management, 3rd Edition,* New York: John Wiley & Sons, 2003.

7

Case Study: B. Stoveburden Trucking Company

Improvement initiatives may be little more than educated guesses at the ideas and activities that have the potential of improving the maturity level of a process or knowledge area. Remember, they may have come as a result of a brainstorming session. Even though they are the task force's educated guesses, they are expected to result in some level of improvement. That expectation may even be stated quantitatively. As a result, we have to continuously monitor the actual improvement and take the necessary steps to continue to deliver improvements until that expectation is met or it is clear that it cannot be met following the current approach. In this chapter we will take a much closer look at the performance of those improvement initiatives and develop action plans to reach specific maturity level goals.

We will illustrate a continuous improvement program, by way of a case study, which is drawn from an actual client engagement (obviously, the names have been changed to protect the client's identity). Not all of the analyses that were conducted produced usable results, so only the more informative analyses will be shown here.

In this chapter we will follow the improvement initiatives from beginning to end. The case study is particularly well suited to this book because it utilizes all of the tools and analyses introduced in the earlier chapters. My purpose in presenting this case study is to demonstrate the power and applicability of the tools. You will also gain a better understanding of how to plan and execute an effective improvement program. I find the case study particularly instructive because it speaks of a situation that I have experienced in a number of client companies. Typical of the client in this case study, as well as several other clients, is that they have established a project management methodology that was the result of a major corporate initiative. Unfortunately, it has not been embraced by the project management community. Many project managers held fast to their old ways because they did not see how the new approach was any better than what they had been using for many years. They were comfortable using their old ways and were not willing to risk taking the new ways. "If it ain't broke, why should I fix it?" was the position that many of them took. In other cases they did not understand the new approach or its documentation and so reverted to their old ways. This is not what the company had in mind, and so they were led to take a more aggressive and thoughtful approach to getting the new methodology integrated into the organization. Does this sound familiar to you? If so, you should find some guidance and sound advice in this chapter. If not, this chapter will prepare you for the day when you do.

7.1 Case Study Background

The B. Stoveburden Trucking Company was formed shortly after WWII by Benny Stoveburden and has run continuously and successfully as a long haul trucker since its beginnings. Its current president is Bea, who is Benny's granddaughter. Benny passed management control of the company to Bea in 2000. Bea had no experience in the trucking business but she did bring some skills that Benny knew would come in handy. Her previous experience included a significant stint as a project manager for a large retail organization in the Midwest. She had personally introduced project management to her employer and has first-hand experience at growing a project management culture from the very beginnings. She knows the value of collaboration in defining, documenting, and implementing a project management methodology and of the bottom line impact that project management can have on an organization. As a result of that experience she championed a similar effort at B. Stoveburden. She appointed a task force of project managers, other managers, and a few senior level managers, and in less than 3 years she had put a project management methodology in place within the IT department (ITD). As part of that effort a PMO had been established as a support for the ITD. The results were less than Bea expected. Usage of the methodology was spotty and the failure rate of projects was unchanged. Bea knew that a more aggressive effort would be needed if the business was to be favorably impacted. She commissioned another task force to investigate the situation. Its membership consisted of project managers. The goal of the project was to reduce quarterly ITD project failure rates from 60% to 30% within 3 years. The failure rate was calculated as follows: of all the IT projects that were begun in a particular quarter, how many failed because:

- They were canceled;
- They were over budget;
- They were completed late;
- They did not meet the client's specifications;
- They were not able to deliver the expected business value.

The percentage was used as the project failure rate for the quarter in which those projects began. As the improvement program put documentation and practice changes in place, the value of this metric would change. The changing values will reflect movement towards or away from the goal failure rate. That movement would direct future initiatives. The current value was 60%. The goal value was to reduce it to 30% within 3 years. If the improvement program was successful, the team would see a continual downward trend as the project failure rate converged towards 30%.

7.1.1 Project Overview Statement

Because this program was so important to the organization, senior management thought that it should be spearheaded by a senior manager. Laurie Driver, the director of the PMO, was chosen as project manager. Laurie was responsible for proposing and selling senior management on the idea of a PMO. She had a lot to gain by the success of this project and she knew it would firmly establish the PMO as a business results organization. She was totally committed to the success of the initiative and had earned the respect of the other senior managers for her work in establishing the project management process they now used.

Laurie began the initiative by assembling a small task force of project managers to help draft the POS. Figure 7.1 is the POS they drafted and had approved by Sal Vation, the CIO.

This POS is short and quite simple. My advice is not to write any more than you need for approval to conduct the project. The more you write, the more reasons you give someone to reject your proposed idea. In this case the impetus for the project came from senior management, not from Laurie. She was simply responding to their request. That means that Laurie did not have to do too much selling to get project approval. The current problem was simply stated and all senior managers would recognize it as a problem. It was no secret that the project failure rate was out of control. The goal was also simply stated: remove the reasons for project failure and bring the failure rate under control. The objective statement could have been more detailed but Laurie felt that that would require some conclusions on her part, and rather than risk the objections of senior management, she chose the low road. The success criterion was given

PROJECT OVERVIEW STATEMENT	Project Name Project Failure Rate Reduction	Project No. PMO.04.02	Project Manager Laurie Driver
Problem/Opportunity The current project failure rate is near 60%, which is unacceptable.			
Goal Design and launch a long-term program to identify and remove the causes of project failures.			
Objectives 1. Identify and categorize the reasons why projects fail. 2. Based on their impact on project success, prioritize the reasons why projects fail. 3. Implement a series of initiatives to reduce the project failure rate.			
Success Criteria The project failure rate will be reduced from 60% to 30% as measured on a quarterly basis within 3 years of the start of this project			
Assumptions, Risks, Obstacles 1. The project failure rate is inversely related to the PD and PP maturity levels. 2. The PMO can effectively enforce project management standards of practice. 3. Senior managers will continuously maintain a high level of support for this project until it is complete.			
Prepared by Laurie Driver	Date 1-14-2004	Approved by Sal Vation	Date 1-16-2004

Figure 7.1 POS for the project failure rate reduction project.

to Laurie by senior management and so there wouldn't be any disagreement here. The assumptions, risks, and obstacles statement seemed plausible.

7.1.2 Fishbone Diagram to Identify the Reasons Why Projects Fail

Before any of the project managers could begin offering their reasons for their projects' failures, Laurie decided to get to the bottom of this situation. She constructed a fishbone diagram from the data of the nine most recent project failures in an attempt to isolate the reason(s) for the failures. If the reasons for the failures could be isolated, she felt that she could initiate a corrective action program to prevent their recurrence. Figure 7.2 is the fishbone diagram the task force supplemented the data collected by interviewing the project managers and teams from the nine failed projects.

As you can see from the figure, Laurie structured her investigation around the five phases of a project. She and her task force of project managers, none of whom had managed any of the nine failed projects, collected the perceived reasons for project failure from each of the nine project teams and integrated that with information gleaned from the project notebooks and the final reports of the nine failed projects. The reasons that they thought their projects failed are listed in the figure. Some of the project managers gave multiple reasons. It is interesting and maybe not surprising that none of the failures are attributed to any actions on the part of the project team. In any case, the data gave Laurie and the team several clues that pointed to four specific knowledge areas where they might focus their attention. They are briefly discussed below.

7.1.2.1 Scope Management

The comments from the project managers clearly point to scope management as an area of concern. Several statements confirm that conclusion:

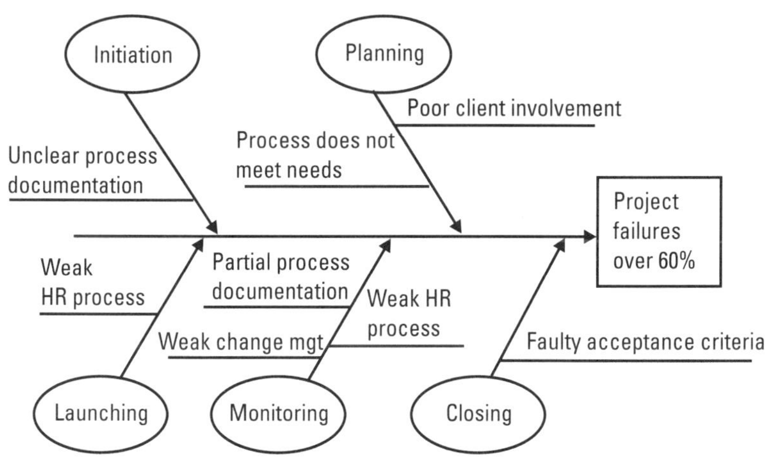

Figure 7.2 Fishbone diagram of reasons for project failures.

- Unclear process documentation;
- Process does not meet needs;
- Weak change management, faulty acceptance criteria;
- Faulty acceptance criteria;
- Poor client involvement.

Keep in mind that there is probably some bias in the project managers' comments. They will tend to defend themselves and point the finger of blame on something other than their performance. Nevertheless, scope management is an area that the task force should investigate further. The data suggests that both the scope management PD and PP maturity levels were suspect. Several improvement initiatives in scope management should probably be commissioned.

7.1.2.2 Human Resources Management

In both the launching and monitoring phases the project managers cited weak HR processes. In their comments they pointed out that HR did not have a sound staffing process or plan in place to help with initial project staffing and with the replacement of team positions that became vacant. Some of the vacancies were the direct result of HR having to move team members from project to project to accommodate higher priority project staffing needs. These factors contributed to several delays in completing task assignments according to the schedule because of the unfilled positions. The teams tried to adapt to the shortages by juggling team assignments but the level of work was overwhelming and schedule slippage was inevitable. More study will be needed to determine what, if anything, can be done to stabilize the staffing situation. It may be that the suspected staff shortage was real and HR was doing its best to accommodate the difficult situation. The reasons given by the project managers may not be accurate.

7.1.2.3 Time and Cost Management

In both the planning and the monitoring phases the project managers expressed concern over the paucity of planning and reporting documentation. What did exist was seen as confusing or too labor intensive to be useful to them and so they often tried to manufacture their own approaches. That clearly was not in keeping with good project management practice and needed to be further investigated by the task force. The project managers may have been right on with their criticism of the documentation. That could be easily confirmed. Assuming that documentation was the problem, the project teams would not be appropriately equipped with tools, templates, and process steps to handle the planning, scheduling, and control of the project. Except for heroic efforts, the project was a high risk for failure.

7.2 PD and PP Maturity Levels for Selected Knowledge Areas

The highest level of interest for process improvement is at the knowledge area level. Individual programs are undertaken for the purpose of improving the maturity level of a single knowledge area from among the nine knowledge areas that define a project management methodology. Several programs can be done concurrently but the risk is high because too much change might be put upon project teams than can be reasonably absorbed. If you exceed the capacity of project teams to accommodate change, the result may be counterproductive. Project failure rates might increase rather than decrease.

A program will consist of several projects as discussed in the previous section. The combined maturity levels of the processes that make up a knowledge area define the maturity level of the knowledge area.

The task force felt the need for more specific data on exactly the size and complexity of the problem they faced and so they asked each of the project managers to complete the scope management, HR management, time management, and cost management portions of the maturity survey for their projects. These were the more significant findings from their interviews of the project managers. Recognizing that there can be some bias in the project managers' opinions, the task force wanted to separate the process problems from the practice problems. The survey would give them the PD and PP data they needed. Figure 7.3 is the result.

Now the task force could isolate their problem even further. There was a major gap between the scope management process definition (PD maturity

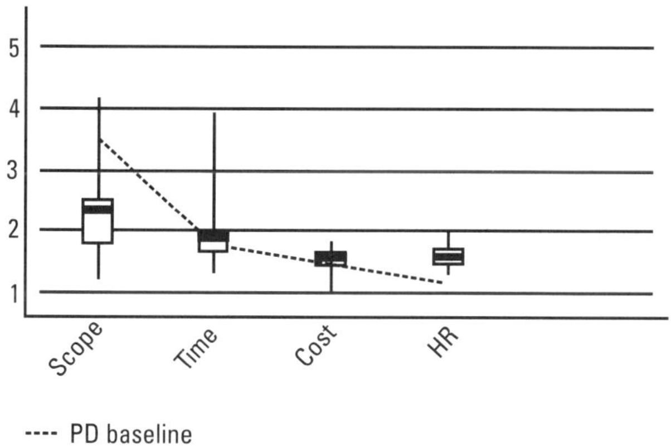

Figure 7.3 Selected knowledge area PD and PP maturity level data.

level) and its practice (PP maturity level). The process definition was at an acceptable level of maturity but the project managers claimed they could not understand it. The data seems to support their opinions. Some teams were using some scope management processes but at a level significantly below the established and documented processes. The reason for that behavior needed further investigation as well.

Finally, there were deficiencies in the project management process definitions for time, cost, and HR management. The project teams were performing consistently up to the level documented in these processes, but that level was not sufficient for good project performance. The consistent performance is clearly evident from the small interquartile range for the time, cost, and HR knowledge areas. Both the PD and PP anomalies had to be corrected.

7.3 Process Level

Now that we have isolated the four knowledge areas that need improvement we have to drill down into each knowledge area to the process level to further refine the focus of our improvement initiatives. At this level of detail we should be able to get at the true causes for project failure. All improvement initiatives are conducted at this level and summarized to the knowledge area level to assure their overall impact. Individual projects are undertaken for the purpose of improving the maturity level of a single process from among the processes that define the knowledge area of interest. This section will target the processes that define the four knowledge areas that surfaced in the prior section: scope management, HR management, time management, and cost management. We will discuss what actually happened in each knowledge area one at a time. In actual practice many of the process improvement initiatives took place concurrently.

7.3.1 Scope Management Processes

Scope management consists of five processes: initiation, scope planning, scope definition, scope verification, and scope change control. The survey data from these five processes is shown in Figure 7.4.

The first observation is that all five processes have large interquartile ranges. That indicates very different levels of process compliance and performance across the teams relative to the PD value of the process. That is due to either different interpretations of the process documentation or the outright avoidance of their use. It is highly likely that some teams decided on the approach they would take to execute a process with little regard for the process documentation. That behavior is not acceptable even if the process

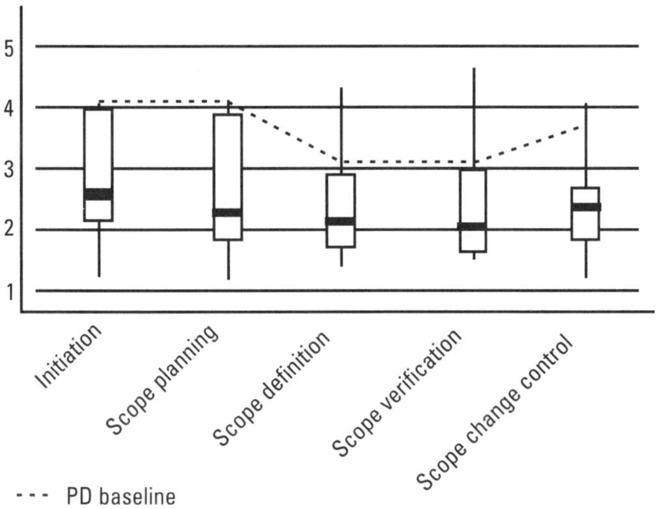

Figure 7.4 Scope management processes PD and PP maturity level data.

documentation is weak, incorrect, or incomplete. If that is the case, the process documentation must be fixed.

There is another explanation for the data. The process documentation may be just fine but the implementation program was lacking. Integrating a new or changed process into an organization is a cultural change. Old habits have to be discarded and replaced with the new. The affected people must see the value in choosing to use the new instead of hanging on to the old. That will not happen just because the new process documentation is published and made available to teams. It will happen because the PMO puts a very carefully designed marketing communications program in place and supports it with training at convenient times and places and in a variety of learning formats. Consulting help from the PMO is another critical ingredient. They need to proactively support teams in the use of the changed processes.

The PD maturity levels of the scope management processes do not indicate a problem with documentation. However, the distribution of PP maturity values does indicate a problem. Note that the interquartile range values are large and extend down from a value close to the PD maturity levels for all but the scope change control process. Also note that the mean maturity values are close to the lower end of the interquartile range. That suggests a definite skew in the distribution of the data points. They tend to collect near the lower end of the range. That suggests that a few teams correctly understood and used the process while most did not understand or use the processes. The observations of the project managers seem to be confirmed. That is, the scope management knowledge area is not clearly documented. While that may not be the only reason for the less than stellar performance of these failed projects, it is a valid partial explanation. If the processes were executed as documented, one would see a much smaller interquartile range. The absence of that property indicates a problem with the documentation.

Scope change control presents a different picture. The fact that only one or two teams followed the documented process suggests that something may be wrong with the process. Laurie was troubled by that result and requested a force field analysis. The results are shown in Figure 7.5.

While the process is documented it appears to be incomplete, burdensome, and not in line with the needs of the project teams. These would seem to be easily overcome by an improvement initiative around the documentation, which should solve the problem. Training is offered but not at a time convenient to teams. Perhaps something other than instructor-led training needs to be offered.

Three initiatives were suggested:

1. Revise the current change management process to be more intuitive.
2. Investigate and remove the causes of the overburden assertion.
3. Implement a more proactive PMO consultant support role to project teams regarding change management and during project reviews.

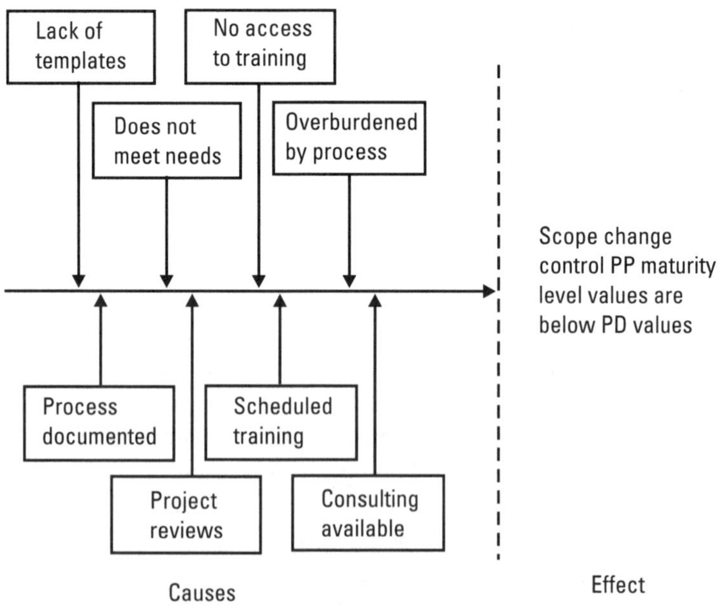

Figure 7.5 Force field analysis of the scope change control process.

The first two initiatives were undertaken and are reported below. The third suggestion did not seem to require an initiative. The PMO was urged to offer support as suggested. Let us take a look at what happened with the first two initiatives.

7.3.1.1 Change Management Process

Because of the results of the force field analysis shown in Figure 7.5, Laurie asked the task force to conduct a root cause analysis of the errors that got into the client change request process. Figure 7.6 is the result. The conclusions are obvious. The change control board (CCB) is dysfunctional. It does not meet its obligation to equitably screen change requests, and it has become a political entity staffed by managers who apparently lack the skill to do their CCB job effectively. It has allowed too many change requests, many of which are not correctly or clearly defined, to reach the project team. This causes needless work by the project team and perhaps that is the reason they feel overburdened. This leads to the conclusion that perhaps the CCB should be disbanded or replaced by some other structure. The other conclusion is that the client does not respect the process but uses it with abandon. They must be made to recognize that a

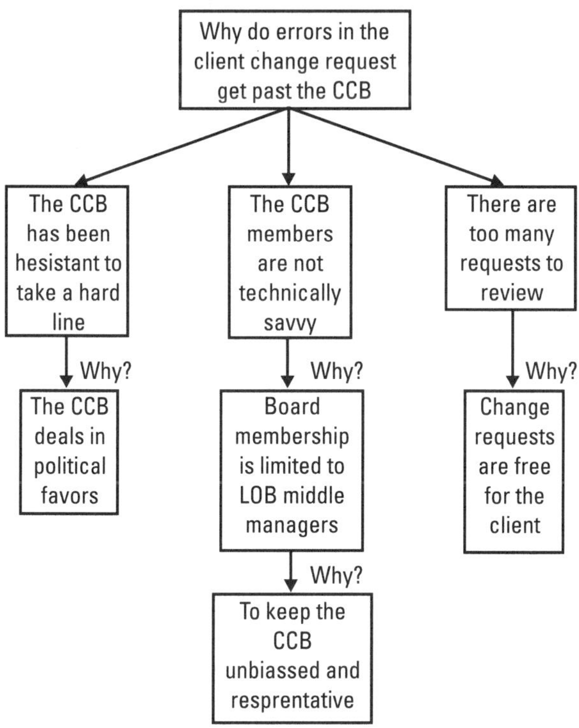

Figure 7.6 Root cause analysis of the errors in the client change request form.

change request takes away team member's time from real project work while they investigate the consequences and value of the impending change. If there is too much of that, it may compromise the project schedule and can certainly be a waste of scarce resources.

One solution is to do away with the CCB and have the client request go directly to the project manager. The project manager should meet with the client to make sure that the request is clearly and completely documented. The project manager may reject the request out of hand and communicate the action and the reason to the client. Alternatively, the project manager may assign one of the team members the task of conducting a project impact study and the appropriate action taken as a result. The current change control process is shown in Figure 7.7.

In this process the CCB is the link between the client and the project manager. If there are errors in the client request, it would seem that the CCB would pick them up and get them corrected before submitting the request to the project team for an impact study. This apparently is not happening and erroneous requests are finding there way to the project team. That requires the team to get

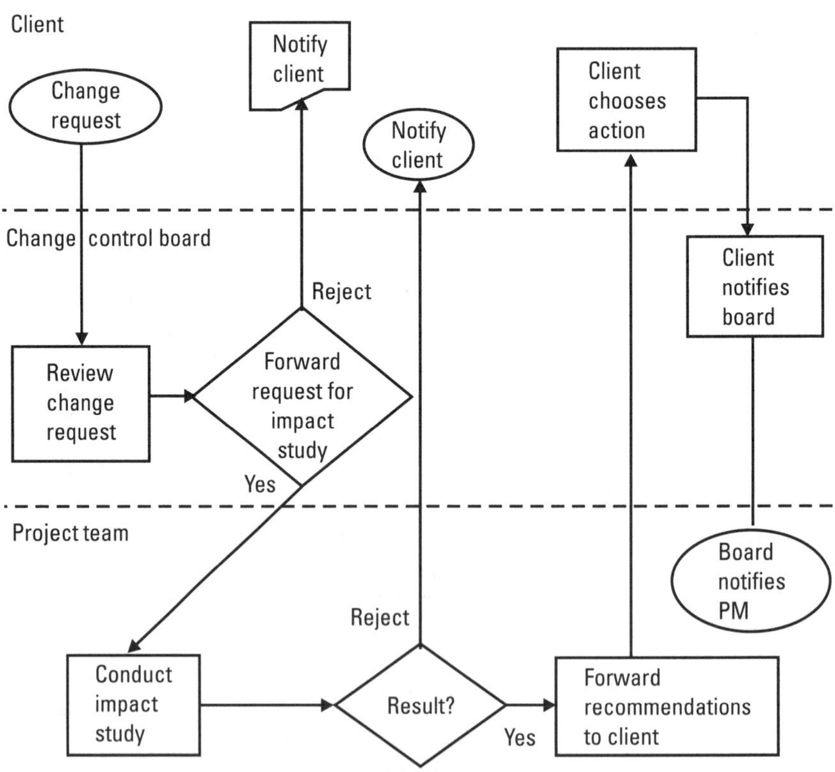

Figure 7.7 Current change management process chart.

clarification from the client. Since the process does not make room for these clarifications, the communications links get muddied and it causes the project team to waste time. Clearly, the review procedures used by the CCB need to be revamped and tightened. When a request for an impact study reaches the project team, the change request must be complete and correct.

The revised scope change management process is shown in Figure 7.8.

The client needs to be made aware of the impact of a change request on the project resources. Frivolous requests require the use of a team member to conduct an investigation and communicate the alternatives. If this is not sufficient to reduce the number of requests, then the client can be charged for submitting change requests. There are several ways to do this that do not destroy a good and harmonious client relationship. A credit may be established as part of the project plan. The credit allows for a specific number of change requests without any payment assessed. After that has been used up, charges will begin for any future requests. Instead of charging the client for a request, a scope bank can be established. At the beginning of the project a credit is established in the scope bank that allows the client a number of prepaid requests. After the credit

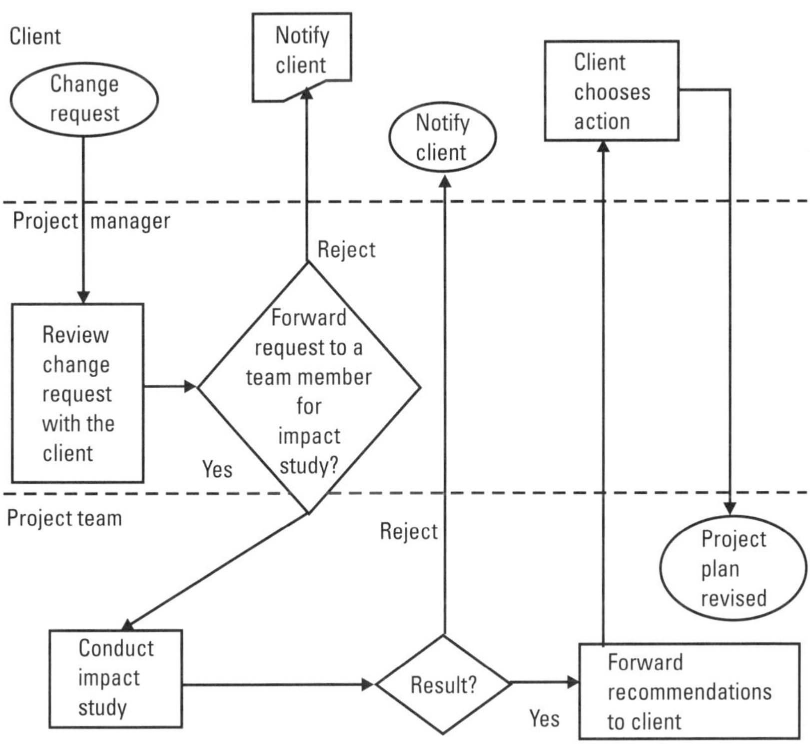

Figure 7.8 A revised change management process chart.

has been used, any further change requests will result in a reprioritization of existing features and functions and a dropping of the low-end priorities to accommodate the new request. Both approaches can be effective. The task force successfully recommended a scope bank with a credit equivalent to 5% of the total task duration.

7.3.1.2 Overburdening Assertion

To test the teams' assertions that they are overburdened by the scope change control process, Laurie reviewed the last 50 situations where there was a change request that required some added attention by the project team. The data is shown in Figure 7.9.

The task force was concerned about the results of the Pareto analysis and decided that the first course of action would be a detailed study of the entire scope change control process. They hoped that that would expose some weaknesses in the current process and suggest ways to remedy all four of the identified causes.

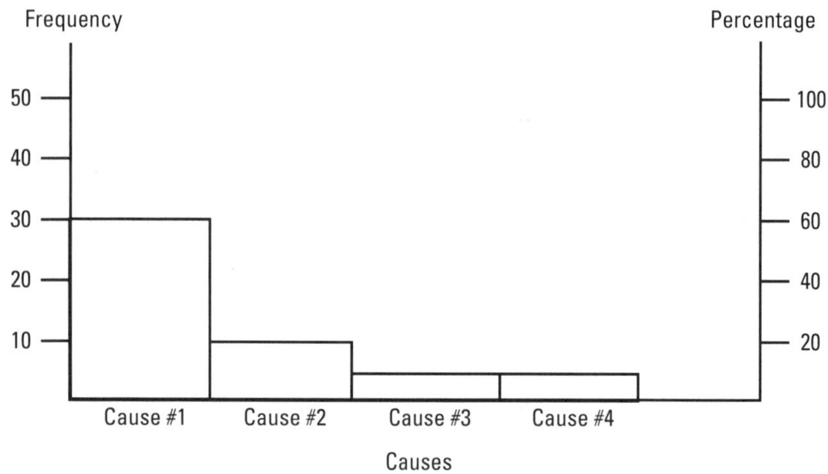

Problem: The project team is overburdened by the scope change control process

Cause #1: Client did not complete change request form correctly and project team must interview them to correct errors.

Cause #2: Client refused to complete the change request form and project team must interview them to get needed information.

Cause #3: There was confusion as to the correct signing authority, resulting in needless transmittal and return of request form.

Cause #4: Change request form did not meet all client needs, resulting in confusion and errors by all parties.

Figure 7.9 Pareto analysis of change requests requiring team attention.

The data shows that the most frequent reason for added attention by the project team was due to the client filing erroneous change request forms; 60% of the added work was due to that reason. A brainstorming session was held to identify possible courses of action. Three ideas surfaced:

1. A more detailed analysis of the specific errors might help in a redesign of the change request form.
2. Client training in the appropriate use of the change request form might also be part of the solution.
3. Do away with the form altogether and substitute an interview process at least for the more significant change requests.

The task force decided that the change request data would be collected during an interview session with the project manager. That would guarantee a correctly completed change request form that could now be confidently forwarded to a team member for analysis. There was no need to change the form or implement client training.

7.3.2 HR Management Processes

HR management consists of three processes: organizational planning, staff acquisition, and team development. The survey data from these three processes is shown in Figure 7.10.

For all intents and purposes it looks like the HR process is not prepared to support project teams. The PP values fall above the very low PD values for all three processes. While the teams may have struggled to establish some of their

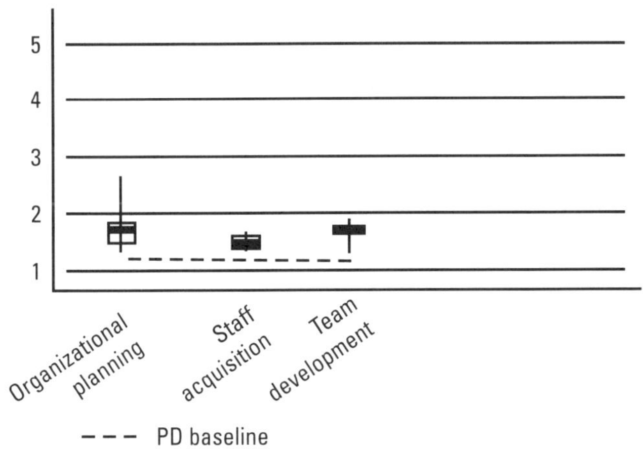

Figure 7.10 HR management processes PD and PP maturity level data.

own HR processes, they were relatively unsuccessful. The first improvement initiative in the HR knowledge area should be to focus on the process definition and documentation to move the PD values near 3.0 for all three processes. Once the necessary changes are in place, the teams should begin to show marked improvement in their PP maturity levels. If not, then a second wave of improvement initiatives should focus on the HR management PP maturity levels.

The delays in filling vacancies on the project teams were disconcerting and needed immediate attention. Because of the staffing problems identified by the project managers, Laurie requested a root cause analysis to investigate why HR could not fill team vacancies in a timely fashion. Figure 7.11 is the result of that analysis.

Two initiatives are suggested by the results:

1. Review the availability report to improve its accuracy.
2. Design and implement a portfolio management process.

7.3.2.1 Availability Report

This initiative uncovered a rather contradictory situation. The project managers were very slow to report availabilities to HR, hence the reason for the delay in

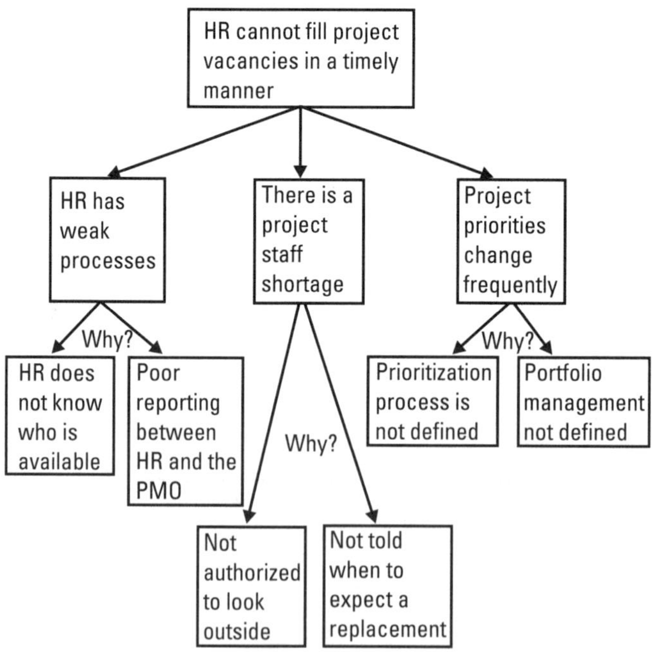

Figure 7.11 Root cause analysis of project staffing vacancies.

reporting availabilities to the project managers. In other words, the project managers were the cause of their own problem, which they incorrectly laid at the feet of the HR department. Apparently, it was not a priority of the project managers to report those availabilities. In order to raise the visibility of the availability report, it was assigned to the PMO. They came up with a data entry and reporting system that made the data available in a timely manner. The PMO now had direct contact with both the contributing project manager and the receiving project manager and that was the impetus needed to get the new system up and running without overburdening the project managers.

7.3.2.2 Portfolio Management Process

The best the task force could discover was that portfolio management was just a step above the "squeaky wheel" stage. There was no repeatable process in place, and in the face of staff shortages one was needed to help decide where the scarce resources should be deployed. Designing and implementing a portfolio management process required a major effort. Another task force was assembled to undertake that initiative. The four-person task force was led by the VP of HR. Laurie and two senior project managers made up the other three members of the task force.

The first task they undertook was to develop a fishbone diagram of the reasons for staff reallocation. Figure 7.12 is the result of that effort. They reasoned

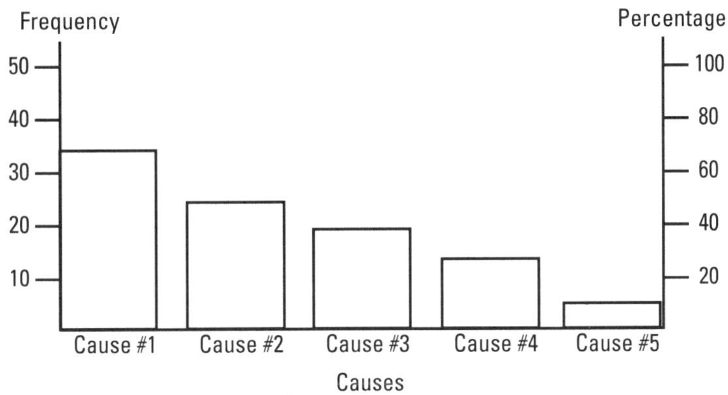

Problem: Team members are being reallocated to other projects and vacancies are not filled in a timely manner.

Cause #1: Insufficient staff resources to cover approved changes.
Cause #2: We have trouble recruiting and hiring skilled IT staff.
Cause #3: Too much executive pressure to staff their pet projects.
Cause #4: Staff resignations are increasing.
Cause #5: The absentee rate is increasing.

Figure 7.12 Pareto analysis of the reasons for staff reallocation.

that if they could identify why staff had to be reallocated, then that would help them design of a portfolio management process where the scarce resource was people not money.

They followed the five phases of the project life cycle.

7.3.3 Time Management

Time management consists of five processes: activity definition, activity sequencing, activity duration estimating, schedule development, and schedule control. The survey data from these five processes is shown in Figure 7.13.

The gap between the PD values and the PP values for the first four processes indicates that the teams are not using the documentation to its fullest. The activity definition process shows the largest PD to PP gap. Because the other processes are dependent on activity definition, there will be problems if activity definition is not done correctly. Correct and complete activity definition is what generates an accurate WBS. All time and cost estimates and scheduling is dependent upon a correct WBS. Without that the project is certain to fail.

The one spurious observation of a PP maturity value at 4.00 was investigated. An industry best practice was uncovered that Laurie felt should be incorporated into the activity definition documentation. The manager of that project was a recent hire and brought a practical "how to" process that he had followed to construct the WBS, which all but assured the team that the WBS would be complete. He offered to document the process and make it available to the PMO for incorporation and general distribution to other project managers. The same project manager offered his approach to activity sequencing that involved use of a whiteboard, sticky notes, and marking pens rather than a software tool.

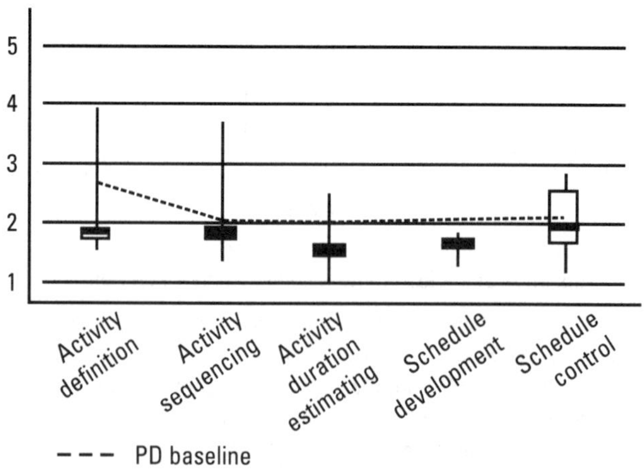

Figure 7.13 Time management processes PD and PP maturity level data.

He also agreed to document that process and contribute it to the PMO for general distribution and incorporation in the activity sequencing process documentation.

There could be several explanations for that behavior and so Laurie requested that a root cause analysis be conducted in an effort to get to the bottom of the problem. Figure 7.14 is the result.

The analysis shed a little more light on the problems associated with the gap between the PD and PP maturity values for the time management knowledge area.

The root cause analysis confirmed the task force's decision to proceed with one initiative. A small working group was appointed to review the activity definition process and recommend additions and clarifications.

7.3.4 Cost Management

Cost management consists of five processes: initiation, scope planning, scope definition, scope verification, and scope change control. The survey data from these five processes is shown in Figure 7.15.

The cost estimating process is interesting. First, the PD value is lower than the PP values of most of the failed projects. Obviously, the teams are using their

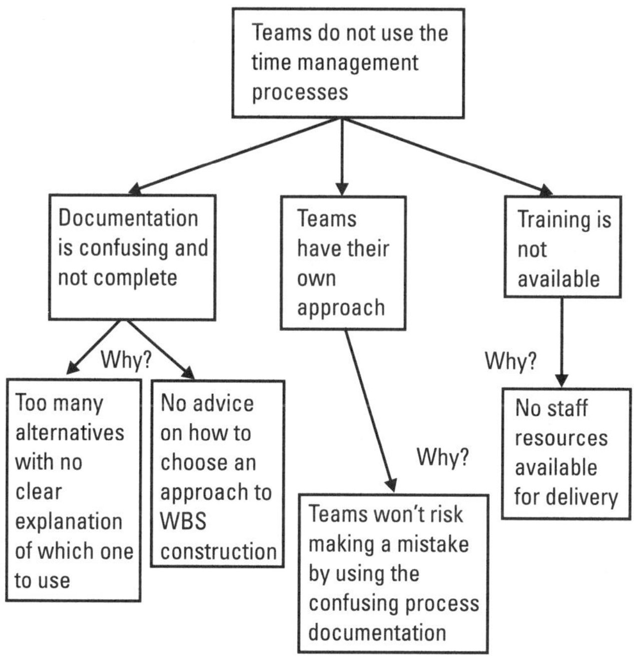

Figure 7.14 Root cause analysis of the documented time management knowledge area.

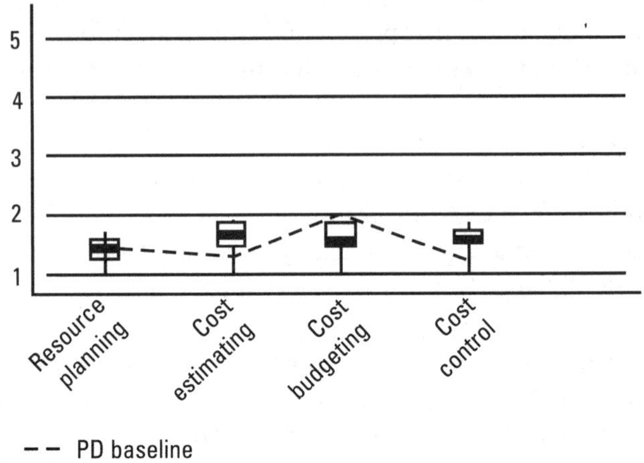

Figure 7.15 Cost management processes PD and PP maturity level data.

own process or have augmented the documented cost estimating process with features and tasks they believe are needed to complete the process. A root cause analysis shed more light on the situation. Figure 7.16 is the result of that analysis.

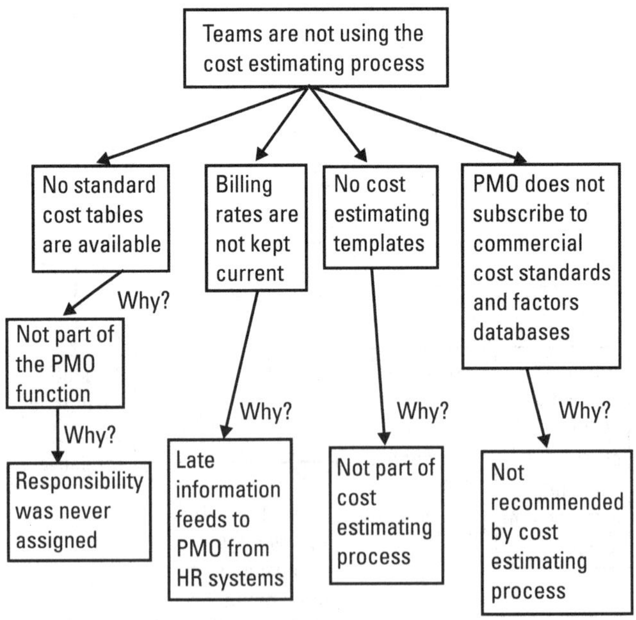

Figure 7.16 Root cause analysis of the cost estimating process.

Three initiatives are suggested by the data:

1. Establish comprehensive cost tables.
2. Establish cost estimating templates.
3. Improve the flow of billing rate information from HR to the PMO.

7.3.4.1 Establish Comprehensive Cost Tables

The PMO was directed to create standard cost tables and a system to collect and distribute the data to teams for planning and budgeting development. The use of commercial databases integrated with the homegrown system was encouraged. The PMO also became responsible for maintaining the currency of the data and responding to requests from the project managers for assistance in the proper use of the data and in the inclusion of additional data elements to facilitate the project budgeting process.

7.3.4.2 Establish Cost Estimating Templates

A long overdue application was suggested by the PMO for the creation of a spreadsheet application that would lead a project manager through the entire process of estimating time and cost. The system would include all the standard costing data in the tables generated from the above initiative. For the first time the organization had a repeatable and reliable method for building time and cost estimates. Under Laurie's direction the IT department designed and programmed the application. It was up and running in less than 3 months. No training was needed as the system prompted the project manager through the entire process. The teams enthusiastically endorsed the application. It was used exclusively by project managers beginning the day it was put into production status.

7.3.4.3 Improve the Flow of Billing Information

As it turns out all three of these initiatives led the task force to the conclusion that the solution was an integrated application that encompassed cost tables, estimating templates, and billing information. IT was given the task of designing such a system. The project was completed nine months later. It has turned out to be a high maintenance application, and while it provides the required information, its future is not certain.

7.4 Results of the Improvement Programs

After 3 years and an aggressive sequence of improvement initiatives, the PD and PP values for the four knowledge areas were as shown in Figure 7.17. To assist

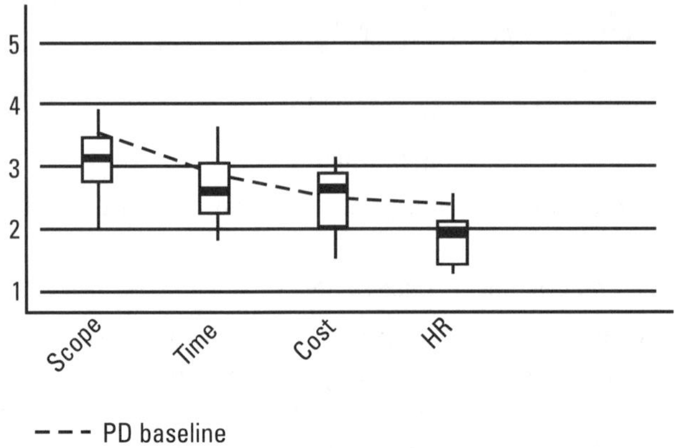

Figure 7.17 Do not overburden project managers and teams with too many process changes at one time.

in the comparison between the before and the after, Figure 7.18 appears as a repeat of Figure 7.3.

The first thing to note is that the PD values for time management, cost management, and HR management have been significantly improved. They are not yet at the targeted maturity levels of 3.0, but they are nearly there. The PP values have kept pace with the PD increases. Note that all process PP averages closely match the PD values except for HR management. This suggests that the implementation and integration of change has been fairly consistent with the PD changes. Senior management was satisfied with these improvements enough to commission a second improvement program under Laurie's leadership and with the support of the same task force. This time the focus would be on two separate programs. The first would be on further improvements in the PP values

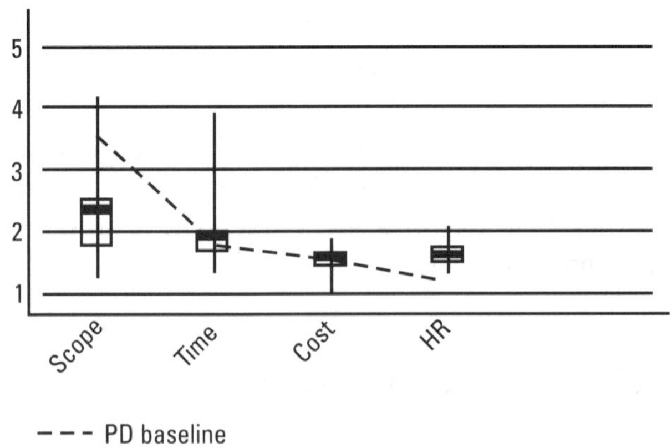

Figure 7.18 PD and PP maturity levels 3 years ago.

for scope management, time management, and cost management. The second would be on the PD and PP values for HR management.

The original goal of the improvement program was to reduce project failures from 60% to 30% in the 3-year period. The actual failure rate moved from 60% 3 years ago to its present level of 42%. Senior management was pleased with that improvement. Although it did not meet their expectations, they were encouraged by the improvement and felt that the second improvement program would result in a 30% or better project failure rate.

7.5 Points to Remember

The following is a list of important points to remember from this chapter:

- Improvement initiatives are exploratory and may be not much better than educated guesses.
- Process changes are also cultural changes.
- Don't overburden project managers and teams with too many process changes at one time.

8

Closing Thoughts

Companies are beginning to realize that they have to get a better return on their project management investment. Many have spent hundreds of thousands of dollars and thousands of hours of employee time building a project management methodology for their organization. They expect to get business value in the way of a higher success rate on the projects they undertake and a more effective and efficient execution of projects. When it does not happen, which is often the case, they need to aggressively develop a strategy to get that return. A continuous quality improvement program centered on project management is their best strategy. Some turn to a portfolio management approach, while others establish a PMO to support projects. Even others invest heavily in a six-sigma program. All of these are admirable initiatives and should be done but, in the end, all of these organizations will find that a continuous quality improvement program for project management is still a necessity.

8.1 Implementation Challenges

A continuous quality improvement program for project management represents a significant change project. Change is always a challenge as it strikes at the very heart of the organization's culture and business processes. Managers and staff have become comfortable in their environment and resist anything that threatens their territory. They are OK with change as long as it does not impact their work patterns or threaten their authority. For project managers it is not any different except that they have the added burden of risk. Any change in procedures brings with it the possibility of increased risk, and project managers tend not to like increased project risk.

8.1.1 Perceived Value

It is WIIFM (what's in it for me). Those who are responsible for establishing and monitoring the project management methodology have a big selling job to do. They have to convince project managers and their teams that what they are promoting as the standard will really make life better for them before they are going to enthusiastically embrace it and use it. Depending on how the standard was put in place, it could either be an easy sell or a nearly impossible sell. Easy if they included those who have to use it on the team that developed it and hard if they did not. I have always thought that the project team would use a process or template if they saw value in it. If they saw no value, no amount of talking or cajoling would change their minds.

8.1.2 Cultural Fit

How well does the project management methodology fit the culture of the organization? I do not recall any of my clients ever asking this question. Is it important that they do? You bet, and here is why. A process-oriented organization will not tolerate an informal approach to project management. That means that all 39 processes must be standardized and thoroughly documented. On the other hand, an informal word of mouth organization that depends on its people to figure out how best to accomplish a task will not tolerate a formal standardized approach. They prefer to let teams figure out what to do. I have had clients like that, and they are often heard saying, "We hire smart people. Put five of them together in a room and they can solve any problem you give them." I suppose this is all right as long as all you have are smart and motivated people. I know a few clients who think they do, but having worked with them I am not convinced they are right.

The caveat here is that you have to keep the organization's culture in mind as you move to change a process.

8.1.3 Sponsorship

Sponsors come and sponsors go. They are usually high-level managers or executives whose half-life can be shorter than your project. As sponsors can change, so can the priority of your improvement initiatives. Add to that the politics of the situation and you have a particularly challenging road ahead of you as you try to manage programs and projects to improve the practice of project management in your organization. If you lose your sponsor and the priority of your improvement program changes, my advice is to stay below the radar with any continuing efforts at process improvement. Continue to pursue improvement initiatives you might want to tackle one process at a time and try to make those improvements through proactive support of project teams rather than through a formal task force approach. This effort will go unnoticed except for the performance reports that show a decrease in project failure rates.

8.2 Suggested Implementation Strategies

I can envision three very different approaches to continuous process improvement. The first, and most aggressive, is the major program initiative. With great fanfare and celebration senior management would convene a task force and challenge them to launch a major offensive to significantly reduce project failures. An aggressive deadline date would be issued and a highly visible and respected senior manager would be appointed task force leader. To me that is like firing a shotgun into an ant hill. You may get a few of them, but chances are you will not get any. The second, the project initiative, is to take a low-key approach and target realistic improvement objectives around the knowledge area that is most in need of improvement. A broad range of authority should be assigned to the manager closest to the problems (probably the PMO director); request a plan with specific goals and time lines. While not as aggressive as the major program initiative approach, it does have a better chance of producing worthwhile results. The third strategy, which I call the slow-but-steady strategy, is to isolate a single cause for project failure and attack it with a targeted project. Obviously, the selected process will be the one identified as a major cause of project failure. This strategy repeats itself until the project failure rate drops to an acceptable level. In the following sections we will explore each of these in a little more detail.

8.2.1 Major Program Initiative

If the project failure rate has reached catastrophic levels and the viability of the organization is threatened, then this strategy may be too late, but it is certainly the only approach that should be considered. Drastic measures are called for and a major program initiative should be commissioned by a C-level manager. For example, the failure rate of development projects in the IT department may be so high that normal business operations are adversely affected and the bottom line is impacted as well. Regardless of why the CIO let the situation deteriorate to that level, immediate and drastic action is needed.

The major program initiative might look something like the B. Stoveburden case study from Chapter 7. An open season is declared on the entire project management methodology environment. That means its documentation and its use are subject to investigation and change. No one is safe from the pervasive eye of the task force!

Imposing a 3-year deadline, as in the case study, may be required because of the business situation, and it may be just the encouragement that the task force needs to deliver the targeted 30% failure rate reduction goal, but it is high risk. There will be a great deal of concurrency with parallel improvement initiatives aimed at all processes in the methodology and the resulting changes coming at project teams from all directions. That will create a lot of short-term confusion as teams try to sort out the changes and integrate them into their work plan. The net result may be the reverse of what was expected. People can only absorb so much change per unit of time. They need time for assimilation and retrospection. The new processes need to become second nature to them and they need to become comfortable in their use and in knowing that the changes really do improve their lot.

The features of the major program initiative are:

- The focus for improvement and implementation is on one process at a time.
- The project management methodology change process is more accommodating and less risk prone than the other approaches.

8.2.2 Project Initiative

This strategy begins much the same as the major program initiative but quickly takes on its own character. In the case study four knowledge areas were identified as in need of both PD and PP maturity level improvements. The major program initiative would have reached the same conclusion. Rather than launch an all-out attack on all four, the project initiative strategy would prioritize the four and work on one at a time. Prioritization could be based on one or more of several criteria, for example:

- Ordered by expected contribution to increased maturity levels;
- Ordered by quick hits by picking the low-hanging fruit first;
- Ordered by major cause of project failure.

This exercise will at least sequence the knowledge areas so that one could be worked on at a time until the expected benefit is attained, and then the next one worked on, and so on until all four knowledge areas have yielded their contribution to increased project success.

The features of the project initiative approach are:

- The entire project community is focused on a single knowledge area. There are no diversions to other knowledge areas as with the major program initiative approach.

- Change will be easier to integrate as compared to the major program initiative approach.

8.2.3 Slow but Steady

Rather than prioritizing knowledge areas, as was the approach in the project initiative strategy, this strategy prioritizes processes. Remember, there are 39 processes to be prioritized. The zone map, which was discussed in Section 3.4 and used in Section 5.7 to prioritize processes for either PD or PP improvement initiatives, is the key to the slow-but-steady approach. Slow but steady has features that make it a good candidate for structuring your improvement program, and you can run this approach without a lot of fanfare and visibility. The features of the slow-but-steady approach are:

- It has minimal staff impact as compared to the other approaches.
- It uses a targeted rifle shot at the most significant process problem.
- It has a better return on investment (ROI) than the other approaches.

8.3 Points to Remember

The following is a list of important points to remember from this chapter:

- Change is always a challenge as it strikes at the very heart of the organization's culture and business processes.
- The project team would use a process or template if they saw value in it. If they saw no value, no amount of talking or cajoling would change their minds.
- A process-oriented organization will not tolerate an informal approach to project management.
- An informal word-of-mouth organization that depends on its people to figure out how best to accomplish a task will not tolerate a formal standardized approach.
- If you lose your sponsor and the priority of your improvement program changes, my advice is to stay below the radar with any continuing efforts at process improvement.
- There are three types of process improvement initiatives:

    1. Major program initiative;
    2. Project initiative;
    3. Slow but steady.

Appendix: Maturity Assessment Questionnaire

The Maturity Assessment Questionnaire consists of more than 800 questions divided into 39 project management processes grouped by knowledge area. Each process consists of a number of questions at each of the five maturity levels. The answers to each question are either "Yes," "No," or "Not Applicable." For ease in identification, the question numbers are composed of three parts. The first character in the number is a letter that denotes the knowledge area. The second digit denotes the maturity level to which the question applies. The final two digits are the question number within that process and maturity level. This codification gives each question a unique number and further identifies the role of the question in the questionnaire.

This version of the questionnaire was developed in collaboration with Babbage Simmel, an Ohio-based training and consulting organization, and is reproduced here with their permission. If you would like more information on how to use this in your organization contact Babbage Simmel at www.babsim.com.

PROJECT INTEGRATION MANAGEMENT

PROJECT PLAN DEVELOPMENT

I101 Do individuals maintain their own versions of the project plan?

I102 Is a scope statement prepared at the discretion of the project manager?

1103 Is a work breakdown structure (WBS) created at the discretion of the project manager?

1104 Are forms and checklists for planning project activities not provided by the organization?

1201 Is there a documented process for developing a project plan?

1202 Is the project plan developed according to the project planning procedure?

1203 Has the organization defined a project life cycle?

1204 Does the organization identify a project sponsor?

1205 Does the project plan have a charter, scope statement and top-level WBS?

1206 Does the project plan include summary-level cost estimates?

1207 Does the project plan include a schedule with major milestones?

1208 Does the project plan include key resource requirements, identified risks, key stakeholders and a communication strategy?

1209 Does the project work plan include staff management items (time reporting, vacation procedures)?

1210 Are project work plans updated to reflect approved scope changes from the change control process?

1211 Do the project manager, project sponsor and appropriate functional managers sign off on the project plan?

1212 Is the project workplan development process fully documented?

1301 Is the project work plan development process fully implemented in all projects?

1302 Does the plan include management plans from cost, schedule, risk, quality, procurement, communications and human resources?

1303 Does the plan include cost estimates and schedule information for visibility and control?

1401 Are all processes in place, documented and being used?

1402 Are project plans incorporated into and support organization strategic plans?

1403 Does data from project plans feed into financial and other organizational systems to complement business execution?

1404 Are best practices and lessons learned captured and made available to other projects?

1501 Is an improvement process in place to continuously improve project plan development?

1502 Are project plans used to support strategic organization decisions and decisions regarding projects?

1503 Is there a process in place to use project plans for strategic organization decisions and decisions regarding projects?

1504 Is the act of planning a project clearly understood?

1505 Is the resource utilization planned?

1506 Are best practices and lessons learned being used to improve planning efforts?

PROJECT PLAN EXECUTION

1105 Is work assigned informally (through verbal communication for example)?

1106 Are deliverables developed informally?

1213 Are status and performance reports produced?

1214 Are technical considerations integrated with cost and schedule information to depict project progress status?

1215 Are basic metrics collected and integrated into project progress reports?

1304 Is summary and detail-level information on deliverables integrated and analyzed?

1305 Are deliverables reports created?

1306 Do status and performance reports include technical information on the project, time spent on project activities and the amount of hours/dollars spent?

1307 Do status or progress reports include informal variance and performance measurement analysis?

1308 Do report templates exist?

1309 Do status and performance reports include information from areas such as scope changes, risk, issues, quality, resources and procurement management?

1310 Are metrics collected from the knowledge areas and integrated into project performance reports?

I405 Are the status and performance reporting processes integrated with the project office, finance/accounting, strategic planning systems and risk management systems?

I406 Are formal variance and performance measurement analyses conducted and reported?

I407 Are best practices and lessons learned captured and made available to other projects?

I507 Is there an improvement process in place to continuously improve project plan execution?

I508 Are the project status and performance reports used to understand the efficiency and effectiveness of a project during execution?

I509 Is the project's overall performance used to support decisions regarding the project and organizational strategy?

I510 Is there a process in place (developed and documented) to use project metrics to support management decisions?

I511 Are best practices and lessons learned being used to improve execution efforts?

INTEGRATED CHANGE CONTROL

I107 Are changes communicated informally to the project manager?

I108 Are changes communicated informally and directly to the team without project manager awareness/involvement?

I109 Are individual project teams applying their own approach to change control?

I110 Are functional, physical and data configuration control informally managed?

I216 Is there a defined and documented change control process for scope changes?

I217 Are scope changes identified with a change request form?

I218 Are scope changes tracked on a change request log?

I219 Are scope changes formally approved?

I220 Are project plans updated incorporating corrective actions?

I221 Is functional, physical and data configuration documented and maintained?

I311 Does the defined and documented project change control system incorporate processes for scope, cost and schedule?

I312 Are the project change control system and processes implemented and used by the project teams?

I313 Is the process repeatable?

I314 Are baselines established and managed to?

I408 Is the project change control process integrated with the organization's control systems, monitoring programs and risk management process?

I409 Is functional, physical and data configuration consistently documented, maintained, managed and controlled for all projects?

I410 Are best practices and lessons learned being captured and made available to other projects?

I512 Are project changes included in the determination of project efficiency and effectiveness?

I513 Is there an improvement process in place to continuously improve the project change control process?

I514 Are historical changes on projects examined to identify trends in change control actions and improve upon the initial project planning process?

I515 Are lessons learned being used to improve monitoring and control efforts?

PROJECT SCOPE MANAGEMENT PROCESSES

INITIATION

S101 Do some projects originate because individuals just decide to do them?

S102 Do some projects originate because they are dictated by management?

S201 Is there a formal project initiation process in place that may be being tested in pilot projects?

S301 Does every project follow a prescribed process for initiation?

S401 Are projects initiated based on their fit within the objectives of the business?

S501 Are there processes in place to continuously evaluate the initiation process and revise it accordingly?

SCOPE PLANNING

S103 Is the procedure used for scope planning at the discretion of the project manager?

S202 Is there a written procedure that describes how scope planning should be done?

S203 Are templates available for identifying and recording scope items?

S204 Is a scope management plan created?

S302 Is the scope management process followed by every project?

S303 Is a scope statement produced for every project?

S411 Is the scope statement justified in terms of higher level business requirements?

S412 Are best practices and lessons learned being captured and made available to other projects?

S516 Are scope statements reviewed for the purposes of improving the scope planning process?

S517 Are best practices and lessons learned being used to improve scope planning?

SCOPE DEFINITION

S104 Is there an informal statement of scope?

S205 Is there a clearly defined and documented process describing the preparation of project charters and scope statements?

S206 Is the project charter and scope statement process enforced by organizational management for large and visible projects?

S207 Is the WBS used as a tool for generating and reporting the scope definition?

S208 Are projects consistently started with the defined project charter?

S209 Are scope statements consistently prepared in accordance with the defined process and format?

S304 Are project assumptions and constraints clearly documented in the scope statement?

S305 Is a statement of work created for each project?

S306 Is the statement of work approved by organizational management?

S307 Is the project scope determined and documented by a fully integrated project team (business unit, technical groups, strategic groups, the client, etc.)?

S412 Are scope statements routinely reviewed against current business requirements and amended accordingly?

S413 Are scope, assumptions, constraints and interproject dependencies thoroughly documented and actively monitored and managed throughout the project?

S415 Are best practices and lessons learned being captured and made available to other projects?

S518 Is the process of determining and documenting scope regularly examined to ascertain process improvements?

S519 Is experience data from a project repository regularly used to improve upon standard templates for scoping and the development of requirements?

S520 Is scope regularly monitored?

S521 Are projected deviations from scope foreseen and carefully documented?

S522 Are projected deviations carefully evaluated based upon value propositions (cost, time, value) before determination is made whether or not to proceed with the deviation?

S523 Are best practices and lessons learned being used to improve scope definition?

SCOPE VERIFICATION

S105 Is scope verification practiced at the discretion of the project manager?

S210 Is the WBS used as input to the scope verification process?

S211 Is there a formal acceptance statement of scope?

S308 Is a formally accepted scope statement required of every project?

S416 Does the approval of a scope statement include a comparison against current business requirements?

S417 Are best practices and lessons learned being captured and made available to other projects?

S524 Is there a formal review process to improve the scope verification process?

S525 Are best practices and lessons learned being used to improve Scope Verification?

SCOPE CHANGE CONTROL

S106 Are changes communicated to the project manager in an informal manner?

S107 Are changes documented in an informal manner?

S212 Is there a defined and documented scope change control process?

S213 Does management support the documented scope change control process?

S214 Does management approve documented scope change on large and visible projects?

S215 For large projects, is there a high degree of compliance in following the scope change control process?

S309 Are the scope change control system, reporting and analysis processes followed by the project teams?

S310 Are scope changes/status being identified, evaluated and managed?

S311 Are stakeholders informed of scope changes/status?

S312 Do you have a way to take corrective action to address scope variances?

S313 Is the performance measurement process repeatable?

S314 Are baselines established and managed to?

S418 Is the scope change control system integrated with the organization's control systems, monitoring programs and risk management process?

S419 Are scope, cost and scheduled reports integrated with technical status reports?

S420 Are best practices and lessons learned being captured and made available to other projects?

S526 In there an improvement process in place to continuously improve the scope control process?

S527 Are metrics gathered and analyzed to ascertain the accuracy of the scoping process?

S528 Are scope variations incorporated into the determination of project efficiency and effectiveness?

S529 Is there a process in place (developed and documented) utilizing scope variances and cost assessments for management decisions during project execution?

S530 Are lessons learned being used to improve monitoring and control efforts?

S531 Are best practices and lessons learned being used to improve Scope Change Control?

PROJECT TIME MANAGEMENT PROCESSES

ACTIVITY DEFINITION

T101 Is the project schedule defined only at the milestone level?

T102 Is activity definition done informally?

T103 Is activity definition documented informally?

T104 Does the activity definition process vary by project?

T105 Is project management software just beginning to be implemented and used to list specific tasks to be performed?

T201 As input to the activity definition process are a scope statements and scope processes prepared?

T202 Are summary activities defined for near-term and long-term efforts?

T203 Is there a basic, documented process for defining activities with standard milestones/exit criteria established for projects?

T204 Are large, highly visible projects using the basic process, WBS template and standard milestones/exit criteria?

T205 Are scope statements prepared as standard practice on large, visible projects?

T206 Does the WBS template go down to at least level 3?

T207 Is the project schedule at a detailed level for large, visible projects?

T208 Are detail-level activities defined (at least to level 3 in the WBS)?

T209 Are detailed activities defined for near-term efforts?

T210 Does the activity definition process include historical information from activities on similar projects?

T211 Is the activity definition process documented and repeatable?

T301 Are scope statements with project assumptions and constraints an organizational standard for all projects?

T302 Is the WBS always used as the basis for determining project activities?

T303 Is a detailed schedule with detailed activities an organizational standard?

T304 Are detailed activities defined for near-term efforts?

T305 Is your project team beginning to identify external, dependent activities?

T306 Does historical information on common activities exist?

T307 Does the activity definition process include activity templates specific to the organization?

T308 Are the activity templates integrated into the standard scheduling software environment?

T309 Are metrics being collected such as the number of activities per project?

T401 Are project activities regularly monitored, focusing on information that is dependent upon other projects or programs throughout the organization?

T402 Does management use project activities to make decisions regarding the project and related efforts?

T403 Is there a decision-making process in place to assist with activity definition?

T404 Are best practices and lessons learned being captured and made available to other projects?

T501 Is a process in place to continuously improve activity definition?

T502 Does the process also focus on ensuring all constraints and assumptions are properly identified and captured?

T503 Are best practices and lessons learned being used to improve Activity Definition?

ACTIVITY SEQUENCING

T106 Are project activities sequenced without the use of a formal process?

T107 Is your project teams on its own when it comes to understanding sequencing methods?

T212 Does the organization have a basic, documented process for sequencing activities and establishing precedence and dependencies?

T213 Does the activity sequencing process include the formal identification of constraints and assumptions that impact the sequencing of activities?

T214 Is the activity sequencing process standard for large, visible projects?

T215 Does the organization have access to different activity sequencing methods?

T216 Are mandatory dependencies identified at a summary level?

T217 Do network diagrams exist at a summary level with mandatory dependencies?

T218 Are discretionary and mandatory dependencies identified at the detailed level?

T310 Is the activity sequencing process expanded to include external dependencies and activity network templates?

T311 Do the network templates depict common, sequenced activities with dependencies?

T312 Is the repeatable process an organizational standard for all projects?

T313 Does your project team document its network diagram approach?

T314 Are unusual aspects of the network diagram documented?

T315 Do network diagrams exist at the detailed level with discretionary, mandatory and external dependencies?

T316 Are the network templates integrated into the standard scheduling software environment?

T317 Is historical information being collected such as the type of external dependencies?

T405 Are project dependencies regularly monitored, focusing on dependencies between projects and programs throughout the organization?

T406 Does management use the dependent relationships to support decisions regarding the project and related efforts?

T407 Is there a process in place that uses project dependencies to understand the full impact of management decisions?

T408 Are best practices and lessons learned being captured and made available to other projects?

T504 Is there an improvement process in place to continuously improve activity sequencing to better identify mandatory, discretionary and external dependencies and to determine when each product or service must be accomplished relative to other activities?

T505 Are best practices and lessons learned used to improve Activity Sequencing?

ACTIVITY DURATION ESTIMATING

T108 Is the estimation of activity duration an informal process?

T219 Are there documented estimation processes?

T220 Are estimated and actual duration results archived for retrieval and use for assisting in the duration estimation process?

T318 Are resource capabilities considered in the duration estimation process?

T319 Are project managers required to use one of the estimation techniques approved by the organization?

T320 Are SMEs routinely involved in the estimation process?

T321 Are alternative estimation techniques documented and available for use?

T409 Is past estimation performance routinely used to analyze current performance?

T410 Are best practices and lessons learned being captured and made available to other projects?

T506 Is there a process in place to review the performance of the estimation process with a view towards improving it?

T507 Are best practices and lessons learned being used to improve Activity Duration Estimating?

SCHEDULE DEVELOPMENT

T109 Is there an informal approach to schedule development?

T110 Is schedule development limited to independent milestones?

T111 Are durations between milestones guesses?

T112 Do project managers have their own way of identifying resources and quantities needed?

T113 Is finding who is available to work on projects done informally?

T221 Is there a full, documented, and repeatable process in place for developing schedules?

T222 Does the process include the development of a historical database to collect data on activity durations?

T223 Is a schedule management plan and process developed and documented?

T224 Do large, visible projects use the process as a standard?

T225 Does the project office closely monitor and support the determination of project activity durations, development of scheduled and establishment of project baselines?

T226 Are project schedules developed at a detailed level?

T227 To calculate durations, do project teams rely upon expert knowledge and access to industry methods, commercial databases and industry standards and factors?

T228 Does the organization have a complete resource listing, industry standard tools, techniques and/or factors for project teams to approximate quantities?

T229 Do teams develop staffing plans and work with line management to acquire resources?

T230 Are the resources inserted into the schedule?

T231 Are cost estimates used to support schedule development?

T232 Are project risks considered in schedule development?

T233 Does the organization have a documented process for allocating, time-phasing and base-lining a project?

T234 Are baselines established?

T235 Are project management software tools standard for large, visible projects?

T236 Are tracks of work integrated within projects?

T237 Does the organization have access to different scheduled methods (deterministic CPM, probabilistic GERT and weighted average PERT)?

T322 Is the project schedule at an appropriate level of detail and in line with the project scope and WBS?

T323 Is a historical database established?

T324 Is the organization collecting and analyzing actual project durations for similar activities?

T325 Is the process to identify resource requirements fully implemented within the organization?

T326 Is the scheduling process fully integrated with the project office, strategic planning systems and risk management processes?

T327 Does the project office centrally manage resource prioritization?

T328 Are baselines established, adhered to and managed?

T329 Are project management software tools standard for all projects?

T330 Are projects integrated within program areas?

T331 Is cost and schedule information integrated?

T332 Are metrics collected and analyzed in areas such as duration standards, capability factors and resource dedication factors?

T333 Are all schedule development processes in place, documented and being used?

T334 Is cost and schedule information integrated with technical information?

T411 Are baseline estimates (revised and original) being used to manage individual projects?

T412 Are baseline estimates (revised and original) being used to make management decisions regarding project execution?

T413 Is resource utilization maximized?

T414 Do variance reports measure performance metrics of efficiency and effectiveness?

T415 Is schedule status used to support management decision-making?

T416 Is there a process in place (developed and documented) using baseline, resource utilization measurements and schedule status for management decisions?

T417 Are best practices and lessons learned being captured and made available to other projects?

T418 Is the baseline process fully integrated with the organization's strategic planning systems and risk management process?

T508 Is an improvement process in place to continuously improve the schedule definition process?

T509 Does the improvement process also focus on the schedule management plan?

T510 Are best practices and lessons learned being used to improve Schedule Development?

SCHEDULE CONTROL

T114 Are individual project teams and segments of the organization applying their own approach to managing and controlling schedules?

T115 Are schedule milestone changes managed differently on a project by project basis?

T116 Are schedule reports provided on as needed basis?

T117 Is schedule performance (metrics) tracked in an informal manner?

T238 Is there a process in place (developed and documented) for managing and controlling schedules?

T239 Has the concept of schedule change control been introduced?

T240 Does the process include items such as schedule statusing, a change control form, a change log and an issues log/form?

T241 Are summary and detailed schedule reports developed and provided to key stakeholders?

T242 Are schedule reports produced from a central system?

T243 Are schedules and statuses tracked using planned versus actuals and milestones complete?

T244 Are schedule baselines established?

T245 Does the organization produce simple variance analysis of schedule status?

T246 Are metrics (schedule baseline, planned status, actual status, etc.) collected?

T335 Are the schedule change control system, schedule reporting and earned value analysis processes followed by project teams?

T336 Are schedule changes/status being identified, evaluated, managed and communicated to stakeholders?

T337 Is a performance measurement process in place (developed and documented) to evaluate project schedule status and take corrective action?

T338 Are schedule baselines established, adhered to and managed?

T339 Are cost and schedule reports integrated?

T340 Are performance metrics (schedule variance, estimates at completion, etc.) monitored and analyzed?

T341 Are corrective actions implemented for performance metrics?

T342 Are all schedule control processes in place, documented and being used?

T343 Is the schedule change control system integrated with the organization's control systems, monitoring programs and risk management process?

T344 Are cost and schedule reports integrated with technical reports?

T419 Are schedule assessments part of the determination of project efficiency and effectiveness?

T420 For certain projects, is earned value and performance reporting integrated with cost and schedule systems?

T421 Does the schedule support earned value analysis?

T422 Does the organization calculate the budgeted cost of work scheduled and performed?

T423 Does the organization calculate the schedule estimate at completion?

T424 Are all earned value techniques used to compare project performance to the baseline and make forecasts?

T425 Are earned value techniques used to update project schedules and to support the determination of project efficiency and effectiveness?

T426 Is there a process in place (developed and documented) that uses schedule assessments and earned value techniques for management decisions during project execution?

T427 Are best practices and lessons learned being captured and made available to other projects?

T511 Is there a process in place to continuously improve the schedule control process including schedule-performance analyses?

T512 Are best practices and lessons learned being used to improve Schedule Control?

PROJECT COST MANAGEMENT

RESOURCE PLANNING

B101 Have project managers developed their own way of identifying resources and quantities needed?

B102 Are functional support areas sometimes overlooked?

B103 Is resource planning done on an informal basis?

B201 Is a complete resource listing defined for all labor categories, equipment and material?

B202 Is everyone encouraged to use the checklist for identifying resources?

B203 Is the checklist standard practice for large, visible projects?

B204 Is a planning process developed and documented to include the resource listing and methodologies for determining quantities?

B205 Does the organization have standard tools, techniques and/or factors for the project teams to approximate quantities?

B206 Is the planning process supported by management and becoming accepted throughout the organization?

B207 Is the generic resource listing incorporated into the project office's resource repository?

B208 Are project-specific requirements manually inserted into the repository?

B301 Is the resource planning process fully implemented within the organization?

B302 Does documentation exist on all planning processes and standards for identifying resource requirements?

B303 Are the project's resource requirements uploaded into the project office's resource repository?

B304 Are metrics collected and analyzed on the types of resources required by projects and the resource availability to determine organizational efficiency in identifying and staffing resources?

B401 Are all resource planning processes in place, documented and being used?

B402 Is the planning process fully integrated with the project office (for resource prioritization and scheduling) and the HR project management process (for resource acquisition, assignment and forecasts)?

B403 Are best practices and lessons learned being captured and made available to other projects?

B501 Is there an improvement process in place to continuously improve resource planning to completely identify all requirements as early as possible and in the right quantities?

B502 Are best practices and lessons learned being used to improve?

B503 Does the planning process include a method to identify an organizational priority for obtaining additional resources during project execution?

B504 Are the project priorities linked to the management decisions?

B505 Can project teams identify the priority of their resource requests?

B506 Is the enhanced process for identifying resource requests developed, documented and in place?

B507 Are lessons learned being used to improve Resource Planning?

COST ESTIMATING

B104 Are estimates developed on an as needed basis?

B105 Is the documentation for the estimates incomplete, limited and not required by the organization?

B106 Do the individual project teams have access to some tools and techniques for cost estimating?

B209 Does the organization have a documented process for generating and documenting project cost estimates?

B210 Has a scope statement been prepared?

B211 Does a top-level WBS template exist?

B212 Is a summary schedule in place?

B213 Has a basic cost-estimating template been established?

B214 Have standard resource billing rates been developed for generic resources?

B215 Does a cost-estimating historical database exist to develop cost standards and factors?

B216 Is there a cost management process in place (developed and documented)?

B217 Is the cost management process standard on all large, visible projects?

B218 Does the organization have access to tools, techniques, commercial databases and industry cost standards and factors?

B219 Does the capability exist to estimate most levels of the WBS resulting in detailed project cost estimates?

B220 Are project risks considered when estimating cost?

B221 Is there a system in place to record project estimates and collect actuals for future comparisons?

B305 Is the cost-estimating process expanded to include cost analysis of alternatives?

B306 Is the entire cost estimating process documented and repeatable?

B307 Is the process outlined within the cost management plan in place and implemented?

B308 Have organization-specific cost standards and factors been developed?

B309 Are comparisons made between actual project costs and the original estimates?

B310 Are metrics collected, analyzed and reported?

B311 Is the historical database established?

B312 Are data collected and analyzed for future reference and quantitative application?

B404 Are all cost estimating processes in place, documented and being used?

B405 Is the cost-estimating process fully integrated with the project office, finance/accounting, strategic planning systems and risk management process?

B406 Do organization-specific cost standards and factors exist for elements of the WBS that are consistently used and standard in projects?

B407 Are best practices and lessons learned being captured and made available to other projects?

B508 Is there an improvement process in place to continuously improve cost estimating to better forecast project costs and improve the cost management plan?

B509 Are best practices and lessons learned being used to improve?

B510 Are comparisons made between forecasted project costs and the original estimates?

B511 Does management use comparisons to historical information to understand required resources for continued support of project activities and to make decisions regarding the project?

B512 Is there a process in place (developed and documented) utilizing cost estimates for making management decisions before and during project execution?

B513 Are lessons learned being used to improve Cost Estimating?

COST BUDGETING

B107 Have project teams adopted their own approach to developing cost baseline?

B108 Is documentation of cost budgeting informal and incomplete?

B222 Is baselining a common practice for large, visible projects?

B223 Does the organization have a documented process for allocating, time-phasing and baselining a project?

B224 Does the project have a staff management plan that supports the development of the time-phased baseline?

B225 Does the capability exist to baseline projects?

B226 Are most projects developing and documenting project baselines at differing levels of detail?

B227 Are baselines established in line with the project schedule?

B313 Are projects developing and documenting project baselines at the lowest reasonable level?

B314 Does the capability exist to enter timephased estimates into the project software environment at any appropriate level of detail?

B315 Is the baseline process fully integrated with the project office's schedule system (or comparable)?

B316 Is the process documented and repeatable?

B317 Are baselines established, adhered to and managed?

B408 Are all cost budgeting processes in place, documented and being used?

B409 Is the baselining process fully integrated with the scheduling, the organization's finance/accounting, strategic planning systems and risk management process?

B410 Are best practices and lessons learned being captured and made available to other projects?

B514 Is there an improvement process in place to continuously improve cost-budgeting and baselining process?

B515 Are best practices and lessons learned being used to improve?

B516 Are baseline estimated being used to manage individual projects?

B517 Are baseline estimated being used to make management decisions regarding project execution?

B518 Is a process in place (developed and documented) utilizing baseline measurements for making management decisions during project execution?

B519 Are lessons learned being used to improve Cost Budgeting?

COST CONTROL

B109 Are individual project teams applying their own approach to managing and controlling costs?

B110 Are cost changes informally managed and not monitored?

B111 Are cost reports provided on a by-request basis?

B112 Is cost performance tracked using nonstandard practices?

B228 Is a process in place (developed and documented) to publish and distribute cost reports?

B229 Are periodic cost reports developed at the summary level and provided to key stakeholders?

B230 Are summary cost reports produced from an integrated system?

B231 Are basic cost metrics (planned budget and percent complete) collected and reported?

B232 Is there a process in place (developed and documented) for managing and controlling cost including cost statusing, a change control form, a change log and an issues log/form?

B233 Has the concept of a cost change control system been introduced?

B234 Have baselines been established in line with the project schedule?

B235 Are summary and detailed cost reports developed and provided to key stakeholders?

B236 Are estimated project actuals provided by the project teams?

B318 Are the cost change control process, cost-reporting process and performance measurement analysis processes followed and used by the project teams?

B319 Are cost changes/status being identified, evaluated, managed and communicated to key stakeholders?

B320 Are baselines established, adhered to and managed?

B321 Are cost and schedule reports integrated?

B322 Is earned value and performance status reporting integrated with cost and schedule systems?

B323 Are performance metrics (schedule variance, cost variance and estimates at completion) monitored and analyzed?

B324 Are corrective actions implemented for performance metrics?

B325 Do the project teams reconcile estimated actuals versus accounting actuals from corporate financial/accounting systems?

B411 Are all cost control processes in place, documented and being used?

B412 Is the cost change control system fully integrated with the organization's control systems, monitoring programs and risk management process?

B413 Are cost and schedule reports integrated with technical status reports?

B414 Are actuals provided by the corporate financial/accounting systems and analyzed by the project teams?

B415 Is there a process in place (developed and documented) to evaluate project cost status and take corrective action?

B416 Is the process documented and repeatable?

B417 Are best practices and lessons learned being captured and made available to other projects?

B520 Is there an improvement process in place to continuously improve the cost control process?

B521 Are best practices and lessons learned being used to improve Cost Control?

B522 Are cost assessments incorporated and included in the determination of project efficiency and effectiveness?

B523 Is there a process in place (developed and documented) for management decisions during project execution?

B524 Are lessons learned being used to improve the monitoring and control efforts?

PROJECT QUALITY MANAGEMENT

QUALITY PLANNING

Q101 Does each project manager develop their own quality plans?

Q201 Has the quality assurance process been enhanced to include processes such as flowcharting, metrics, or quality control measures?

Q202 Do metrics include results of reviews and tests against criteria, specifications, quality standards and business requirements?

Q203 Is the quality planning process considered the standard way of ensuring quality is accounted for within the project's products/services?

Q204 Are most projects using quality planning processes, including the development of a quality management plan?

Q205 Are all large projects using quality planning processes, including the development of a quality management plan?

Q206 Does management sign off on the quality plans for larger projects?

Q301 Does the quality planning process include guidelines for design of experiments (analytical techniques that help identify which variables have the most influence on the overall outcome)?

Q302 Does the quality planning process include emphasis on quality milestones?

Q303 Does the quality planning process include standardized checklists for the use of the project teams in creating their quality plans?

Q304 Does the process include a formal quality plan?

Q305 Does the process include templates to create a formal quality plan?

Q306 Does the process include organizational management at key approval points?

Q307 Does the quality planning process include the scope/perspective of other entities in the immediate domain of the products of the project?

Q308 Has the organization identified one or two people whose focus is organizational project quality standards and assurance?

Q401 Does the quality planning process include the perspective of the entire environment into which the product of the project is being placed?

Q402 Is a quality office established within the organization?

Q403 Does the organization benchmark its project results against industry standard results?

Q404 Are best practices and lessons learned being captured and made available to other projects?

Q501 Does the quality planning process include a process whereby the process itself is critiqued throughout the project?

Q502 Are best practices and lessons learned being used to improve Quality Planning?

Q503 Does the quality planning process integrate the usage of metrics and data from earlier projects, and from earlier stages of the given project, to make benefit/cost comparisons and decisions regarding quality planning?

Q504 Does the quality planning process use metrics and data collected to make value-related trade-off decisions regarding quality planning?

QUALITY ASSURANCE

Q102 Do some project teams establish their own quality assurance procedures for their projects to make sure everyone is following the same procedures?

Q207 Is there a documented approach for quality assurance?

Q208 For large, highly visible projects, do teams establish project procedures and use walkthroughs or peer reviews to assure the team is following procedures?

Q208 Does the team identify the points in the development process at which there may be need for extra quality precautions?

Q209 Are quality assurance procedures, including tools and techniques such as flowcharting and operational definitions, considered standard approaches on large, highly visible projects?

Q210 Does the project office provide a quality policy and project management processes and standards to support the quality assurance measures on a project?

Q211 Have project teams devised checklists for use in checking/promoting quality throughout the project life cycle?

Q212 Do some project teams use tools and techniques, such as Pareto Charts, Cause and Effect Diagrams, Force Field Analysis, Decision Trees or Control Charts throughout the project?

Q309 Are tools and techniques (Pareto Charts, Cause and Effect Diagrams, Force Field Analysis, Decision Trees or Control Charts) considered standard approaches on large, highly visible projects?

Q310 Are tools and techniques (Pareto Charts, Cause and Effect Diagrams, Force Field Analysis, Decision Trees or Control Charts) suggested approaches on all projects?

Q311 Do projects take on a proactive stance in quality and plans for regular walkthroughs with organizational management to assure that the product will function correctly with the other elements of the environment?

Q405 Do projects regularly include walkthroughs with organizational management and other project teams as major portions of the product are developed to assure that the product will meet the business requirements and specifications?

Q406 Are documentation procedures in place that require each major component of the product to be fully documented prior to moving into a final production mode?

Q407 Do nearly all of the projects use quality assurance processes/methods as specified by the quality standards?

Q408 Are metrics regularly collected throughout the project life cycle and compared to industry standards to ascertain potential problem areas?

Q409 Are best practices and lessons learned being captured and made available to other projects?

Q505 Is feedback gained from the quality assurance processes and used to improve project management processes for future projects?

Q506 Are effectiveness and efficiency of the product and project regularly measured using metrics collected throughout the project?

Q507 Are best practices and lessons learned being used to improve Quality Assurance?

QUALITY CONTROL

Q103 Do project team members seldom have someone look over their work before they submit it to management?

Q104 Does informal testing occur on specific units or portions of the product for development-based projects?

Q213 Does the basic quality assurance process include suggested approaches for quality control (like guidelines for testing or review of individual project deliverables by the project team)?

Q214 Are summary-level testing metrics collected and evaluated?

Q215 Do the tools used include acceptance criteria, performance standards, business requirements, specifications and quality standards for review/testing?

Q216 Are the quality control processes used on large and highly visible projects?

Q217 Are the quality control processes encouraged on all other projects?

Q218 Are full acceptance criteria and specifications developed?

Q219 Does the quality process include templates, guidelines and test plans for deliverable and product testing, including evaluation by the customer?

Q220 Does the quality process include templates, guidelines and test plans for unit testing (testing of the individual components of the product)?

Q221 Does the quality process include templates, guidelines and test plans for integration testing (testing how the major assemblies of the product work together)?

Q222 Does the project team do the unit and integration testing?

Q223 Are summary-level and detailed testing metrics collected and evaluated?

Q312 Are project performance standards identified and measured against?

Q313 Does the quality process allow for product testing in a simulated situation?

Q314 Does the quality process include templates, guidelines and test plans for integration testing (testing how the product works in its environment with other systems/products)?

Q315 Does the quality process include templates and guidelines for acceptance testing (testing the human interface and testing against the original requirements)?

Q316 Is the client actively involved in interface testing?

Q317 Does the project team perform the actual interface testing?

Q318 Does the client drive acceptance testing?

Q319 Does acceptance testing require signoff from both the client and management before being accepted as complete?

Q409 Does the organization have performance standards in place?

Q410 Are project products consistently tested against performance standards?

Q411 Does the organization have functional standards in place?

Q412 Are project products consistently tested against functional standards?

Q413 Does the quality process include templates and guidelines for testing that the products integrate with other products/systems in the organization, and that all portions of the organization affected in any way have an opportunity to adequately test those effects?

Q414 Are best practices and lessons learned being captured and made available to other projects?

Q508 Are quality control/testing results regularly examined throughout the project to ascertain improvements?

Q509 Are quality control/testing results integrated back into the process?

Q510 Does management use quality control results to make decisions on the usability and fit of the product and when/if the product is acceptable?

Q511 Are best practices and lessons learned being used to improve Quality Control?

PROJECT HUMAN RESOURCES MANAGEMENT

ORGANIZATIONAL PLANNING

H101 Are project managers assigned on an ad hoc basis?

H102 Is there an informal approach to determining how many people are required to work on project activities?

H103 Does an informal reporting structure exist such that project staff members know that they need to get their assignments from the project manager?

H104 Is an informal analysis conducted to define organizational, technical and interpersonal interfaces that exist within the organization?

H201 Does the project manager create a basic overview of the types of skill sets that are required by the project and the approximate timeframe in which these skill sets are needed?

H202 Does a basic responsibility definition exist in the form of a responsibility assignment matrix by major deliverable?

H203 Does a project organization chart exist so that the individuals on the project know who reports to whom on the project?

H204 Is there an understanding of the constraints that may be prevalent in attaining required resources, such as the type of organization and individual preferences to work on one project or another?

H205 Is there a narrative description of the responsibilities for the key project personnel and a staffing plan that defines when resources will be needed?

H206 As the project progresses, is there a measurement of planning versus actual that occurs with regard to the staffing plan?

H207 Does the staffing plan provide corrective action?

H301 Is a formal analysis conducted to define the organizational, technical and interpersonal interfaces that exist within the organization?

H302 Are constraints that may be prevalent in attaining required resources analyzed and a response developed?

H303 Is there a narrative description of the responsibilities for all project personnel?

H401 Is project organizational planning integrated into the overall resource pool management and prioritization?

H402 Is an action plan developed to deal with the organizational, technical and interpersonal interfaces that exist within the organization?

H403 Are constraints to resource planning managed?

H404 Is there commitment by all stakeholders to the definition of the roles and responsibilities in the staffing plan?

H405 Does integrated decision-making (decisions are evaluated based on their impact to the project and the organization) begin to occur?

H406 Are best practices and lessons learned being captured and made available to other projects?

H501 Is organizational planning evaluated on a periodic basis?

H502 Are enhancements to the process continuously incorporated?

H503 Are performance metrics for HR used to define efficiency and effectiveness of resource utilization throughout the project?

H504 Is stakeholder analysis effectiveness and efficiency evaluated to ensure continuous involvement and signoff throughout the project?

H505 Does integrated decision-making (whereby all decisions are evaluated with regard to their impact on other projects) occur in all projects?

H506 Are best practices and lessons learned used to imrpove Organizational Planning?

STAFF ACQUISITION

H105 Is there an informal approach to finding who is available to work on project activities?

H106 Is there an informal approach to asking line management for having certain resources for a project?

H208 Does staff acquisition consist of identifying the individuals who have the requisite skill sets and time availability to work on the project?

H209 Does the project manager request line management to reserve team members for a certain timeframe?

H210 Is a staffing requirements document submitted from cost management as an input for defining the staff management plan?

H211 Is there a "first come, first serve" process in place whereby whoever requests a resource first gets usage of that resource first?

H212 When a resource is assigned to a project, does line management document the resource's labor category, so that the project manager can use the information for costing purposes?

H213 Is there a process in place that allows organizational management to temporarily reallocate resources if another project or assignment of extremely high priority comes up?

H214 Does the staffing plan include defining the parameters for the desired project team, including minimal experience, personal interests and characteristics, and availability to determine a good fit among project team members?

H215 Does the project manager accept whatever resources are assigned by the line manager?

H216 Do project HR management and the project office coordinate efforts in resource pool management?

H304 Does the project manager work with the project office and line manager in resource pool management and prioritization?

H305 Does the project manager negotiate with line management or get "preassigned" for specific resources?

H306 Upon occasion, does the project manager look outside the organization for specific resources?

H407 Does the project office have an effective resource pool management and prioritization process in place that is used by the line and project managers in fulfilling project resource needs?

H408 Are resource variance reports developed for all projects?

H409 Are best practices and lessons learned being captured and made available to other projects?

H507 Is enterprise resource forecasting evaluated for continuous improvement and enhancements?

H508 Are project manager resource requests evaluated against the resource pool constraints and prioritization to ensure maximization of resource utilization effectiveness and efficiency?

H509 Do resource variance reports measure performance metrics of efficiency and effectiveness?

H510 Is enterprise resource forecasting being used?

H511 Are best practices and lessons learned being used to improve Staff Acquisition?

TEAM DEVELOPMENT

H107 Is there an informal approach to trying to ensure that project team members work together in a professional manner, which may include occasionally trying to get complementary personalities on the same project team?

H108 Is there an informal approach to team meetings whereby the team may be included in an explanation of the project organization, deliverables, scope, WBS and the like?

H217 Are projects begun with an informal kickoff in which the team members are briefed on the purpose of the project, their responsibility and introductions to each other?

H218 Is there a specified process for incorporating the team into scope development and the development of work plans?

H219 Are there guidelines in place for project initiation team meetings, scheduled status reviews, business reviews, technical reviews and a plan for regular and ongoing project reviews?

H220 Do the project initiation meetings and reviews include the team?

H221 Are regular status and progress meetings conducted to keep project team members apprised on how the project is progressing as well as deal with issues that may arise?

H222 Does the project manager contribute to the performance evaluation of the individual team members?

H223 Is a rewards and recognition system established?

H224 Has a conflict management process been defined?

H225 Does management enforce the process for team buy-in to ensure that teams are actively involved and integrated into scope planning and management of the project?

H226 Is a staff development plan developed with the organization responsible for the professional development initiatives?

H307 Does the project manager work with the project office and line management to establish communication?

H308 Are project team peer evaluations conducted by the team for individual peer performance?

H309 Is there a documented conflict management process?

H310 Is the conflict management process being used on all projects?

H311 Does management participate in the team buy-in process?

H312 Does a fully integrated project team include the business unit, technical groups, strategic groups, the client, etc.?

H313 Is all stakeholder input solicited and incorporated into project planning and execution?

H410 Does the organization adhere to a team development process to foster team concepts throughout the organization?

H411 Is a team development process established by which teams on medium and larger projects are expected to evolve?

H412 Are team member training needs identified and communicated to the project office and line management?

H413 Does the project office and line management proactively work with team members to meet their training needs?

H414 Does the project manager significantly contribute to the performance evaluation of the individual?

H415 Is the team buy-in process engaged and used by the majority of projects?

H416 Is management actively engaged in the team?

H417 Is the project board (representing all stakeholders) actively involved in the project on a regular basis?

H418 Are best practices and lessons learned being captured and made available to other projects?

H512 Does the organization value investing in its people throughout the organization?

H513 Does the organization actively ensure that project teams have all that is required to succeed on a regular basis?

H514 Is this question continually asked: Are there ways in which we could get better team buy-in?

H515 Is the information gathered from that question used to improve the overall process?

H516 Are team member training needs forecasted and acknowledged as a value-added investment for the organization?

H517 Is a project conflict management process integrated into the overall corporate management system and efficiency and effectiveness measurements gathered?

H518 Is team satisfaction measured?

H519 Are best practices and lessons learned being used to improve Team Development?

PROJECT COMMUNICATIONS MANAGEMENT

COMMUNICATIONS PLANNING

C101 Does the project manager provide status only when requested?

C201 Is an informal stakeholder analysis developed whereby project stakeholders are identified and provided project summary reports?

C202 Are project constraints and assumptions defined?

C203 Does management encourage large and highly visible projects to provide summary reports at periodic intervals through the expected life of the project?

C204 Is a formal stakeholder analysis conducted and communication requirements defined by combining the type and format of information required with an analysis of the value of that information?

C205 Is a communications management plan developed by large and highly visible projects?

C206 Is a communications management plan encouraged by all projects?

C207 Does the communication plan identify the communication needs from project commencement to closure, including the postproject review and lessons learned?

C208 Is the immediacy of the need for information analyzed?

C209 Is the availability of technology analyzed?

C210 Is the expected project staffing analyzed?

C211 Is the schedule of the project analyzed?

C212 Are project risks analyzed?

C301 Is there a communications plan for all projects?

C401 Is there a method for updating and refining the communications management plan as the project progresses, develops and is incorporated into the corporate systems?

C402 Are best practices and lessons learned being captured and made available to other projects?

C501 Is communications-planning documentation analyzed for value-added impact?

C502 Are best practices and lessons learned being used to improve Communications Planning?

C503 Is communications planning tightly linked with organizational planning?

INFORMATION DISTRIBUTION

C102 Is information distributed on an as needed basis such as a response to a specific request or question about the project directed to the project manager?

C213 Is information distributed via electronic medium or hard-copy documentation?

C214 Is information hand-delivered or mailed to the stakeholders?

C215 Is a basic retrieval and distribution process in place?

C216 Is effective interteam communication of project actuals established?

C217 Are project stakeholders directed to a specific shared file on a computer network or central physical location, where they can retrieve needed project information?

C218 Is the project manager responsible for ensuring that project information is retrieved in a timely fashion and that the stakeholders obtain the information that they need?

C302 Is there a formal information retrieval system by which project stakeholders can retrieve information through an electronic database or central repository (physical or electronic)?

C303 Is there a formal information distribution system including project meetings, hard copy documentation, shared access to networked electronic databases, fax, email and voice mail?

C403 Is there an automated information retrieval system in place that is based upon a database structure and inquiry process?

C404 Does the information distribution system include meetings of varying formats and also multimedia distribution (intranet, internet, and video conferencing)?

C405 Are best practices and lessons learned being captured and made available to other projects?

C504 Is documentation for the entire project available for review during administrative closure?

C505 Are best practices and lessons learned being used to improve Information Distribution?

C506 Are project stakeholders educated and capable of accessing any project-related information that they need in a timely fashion?

C507 Are lessons learned about effective information retrieval and distribution collected?

PERFORMANCE REPORTING

C103 Can informal reporting about the current status of the project be obtained from the project manager?

C219 Can three types of summary reports (for status, progress and phase completion) be produced throughout the life of the project at periodic intervals of project timeline?

C220 Do these reports track milestone attainment of scheduled items?

C221 Has the project been baselined?

C222 Can actuals be collected?

C223 Do project progress reports provide information on schedule, cost and project performance?

C224 Is a formal acceptance document being used for the customer to acknowledge acceptance or project deliverables?

C225 Is there a formal customer signoff at the conclusion of the project?

C304 Are graphical performance reporting chart used such as S-curves (indicate how the funds or hours are being spent), histograms and tables?

C305 Is narrative reporting of a project status being used?

C306 Is informal variance/trend analysis conducted comparing the actual project results to the planned results?

C307 Is informal trend analysis done to determine the estimate-to-complete for budget and schedule parameters?

C308 Is management involved in the identification, analysis, approval (or not) of changes to the project plan?

C309 Are performance reviews conducted to assess project status or progress?

C310 Are project reports archived for future reference?

C406 Do all projects capture performance measurements (such as earned value) for understanding and analysis of project performance?

C407 Is formal variance/trend analysis conducted on projects?

C408 Are best practices and lessons learned being captured and made available to other projects?

C508 Are best practices and lessons learned being used to improve Performance Reporting?

C509 Are the results from the lessons learned imbedded back into the process for continuous enhancements?

C510 Are performance metrics used to define efficiency and effectiveness metrics for projects?

C511 Are lessons learned on all projects captured for future reference?

ADMINISTRATIVE CLOSURE

C104 Are final project reports done at the discretion of the project manager?

C226 Is there a template for the final project report?

C227 Is there a formal sign-off by the client that the project is complete?

C228 Is there a process for capturing and archiving lessons learned?

C311 Is there a documented project closing process that all projects are following?

C312 Are final project reports required and archived for future use?

C409 Are best practices and lessons learned being captured and made available to other projects?

C410 Is there a required process for validating that the project success criteria have been met?

C512 Is administrative closure continuously reviewed with a view towards improvements?

C513 Are best practices and lessons learned being used to improve Administrative Closure?

PROJECT RISK MANAGEMENT PROCESSES

RISK MANAGEMENT PLANNING

R101 Is risk planning left to the discretion of the project team?

R102 Is risk planning an informal activity?

R201 The organization has a documented risk management policy?

R202 Are there risk management planning templates?

R203 Has a risk planning process been defined for the organization?

R301 Is a formal risk management analysis required on all projects as part of the initial approval process?

R401 Is there a formal risk review process?

R402 Is project risk used to compare and prioritize projects in the organization's project portfolio?

R403 Are best practices and lessons learned being captured and made available to other projects?

R501 Is the risk management process continuously reviewed with a view towards improvements?

R502 Are best practices and lessons learned being used to improve Risk Management Planning?

RISK IDENTIFICATION

R103 Do project team members rarely suggest potential risks to the management or stakeholders?

R104 Do risk discussions typically take place when the risk is already a current problem versus a future possibility?

R105 Do project managers informally use scope statement, WBS, project milestones and deliverables to help identify project risks?

R106 Are project team discussions on risk sporadic and informal?

R204 Does the organization have a documented process for identifying project risks?

R205 Is the process encouraged for all projects?

R206 Is the process considered standard for large, highly visible projects?

R207 Is a conscious effort made to identify total project risks (near-term and longer-term) in as much detail as makes sense?

R208 Do risk discussions include input from key stakeholders?

R209 To help identify risks, does the project team have a scope statement and a WBS template that goes down to at least level 3, a more detailed project schedule and a more comprehensive project cost estimate?

R210 Does the project team examine the procurement management plan and staff management plan to help identify risks?

R211 Do risk discussions typically focus on project scope, schedule, and cost risks?

R212 Are top-level risks included in the project team project plan?

R213 Does the project team rely upon expert judgment and known industry lessons learned to identify risks?

R302 Does the organization have a documented repeatable process for identifying project risks which is fully implemented?

R303 Does documentation exist on all processes and standards for identifying risk events?

R304 Does the process include efficient avenues for teams to identify risks (checklists, automated forms, etc.)?

R305 Does your team identify symptoms of risk (risk triggers) for incorporation into the historical database for risks?

R306 Is a conscious effort made to identify total project risks and program risks (interrelationship among related projects)?

R307 Do risk discussions include input from past, similar projects, industry lessons learned and key stakeholders?

R308 Are risk information and symptoms consolidated and integrated?

R404 Are all processes in place, documented and being used?

R405 Is the risk identification process fully integrated with cost management and time management processes and the project office?

R406 Is a conscious effort made to identify total project risks (within individual projects, within programs and between projects/programs)?

R407 Are risks identified with the organization and project in mind?

R408 Are best practices and lessons learned being captured and made available to other projects?

R503 Is there an improvement process in place to continuously improve risk identification to completely identify all risks as early as possible?

R504 Are best practices and lessons learned being used to improve Risk Identification?

R505 Does the risk identification process include a method to identify an organizational priority for the project?

R506 Does overall risk exposure influence management decisions?

R507 Does overall risk exposure give project teams the ability to identify the priority of their risk concerns?

QUALITATIVE RISK ANALYSIS

R107 Is qualitative risk analysis done at the discretion of the project team?

R214 Are there established qualitative risk analysis methods and tools?

R215 Does a risk matrix that defines probability and impact exist?

R216 Can a prioritized list of risks be created?

R309 Are all projects required to create a risk matrix?

R409 Is there an overall risk ranking for the project?

R410 Are best practices and lessons learned being captured and made available to other projects?

R508 Is there a process in place for the continuous improvement of the qualitative risk management process?

R509 Are best practices and lessons learned being used to improve Qualitative Risk Analysis?

QUANTITATIVE RISK ANALYSIS

R108 Does the project manager speculate on the impact to the project if a risk should occur?

R109 Are risk discussions done without analysis, forethought or standard approach/process?

R217 Is a process in place and documented which provides a structured approach to quantifying risks?

R218 Does the process include a standard methodology that will ensure the organization consistently assesses the risk items?

R219 Does the common methodology include low-medium-high ratings or expected monetary value of risks using simple probability and value calculations?

R220 Do project teams attempt to employ more objective approaches to quantify the probability that a risk event will occur and the significance of the impact if it does occur?

R221 Are risks evaluated on a project-by-project basis?

R222 Are risks prioritized based upon a single factor such as monetary value considerations?

R310 Does the process include more advanced procedures for quantifying risks (methodologies such as range predictions, optimal calculations using simulation tools and decision trees and weighted average calculations)?

R311 Does the process include multiple criteria to prioritize risk items (expected monetary value, criticality, timing and risk type)?

R312 Is the entire process documented and repeatable?

R313 Are risks evaluated on a program/organizational basis?

R411 Are all processes in place, documented and being used?

R412 Is the risk quantification process fully integrated with cost management, time management, finance/accounting and strategic planning processes, and the project office?

R413 Does risk quantification take into effect the risks on other projects and other parts of the organization?

R414 Are risks evaluated on an organizational basis?

R415 Are performance indexes included in calculating the impact of risk on a project?

R416 Are best practices and lessons learned being captured and made available to other projects?

R510 Is there an improvement process in place to continuously improve risk quantification to better quantify risks and adequately capture the cost and schedule impact?

R511 Are best practices and lessons learned being used to improve Quantitative Risk Analysis?

R512 Does management use the quantified risks to make decisions regarding the project?

R513 Is there a process in place (developed and documented) that uses absolute values of risk for making management decisions before and during project execution?

R514 Are lessons learned being used to improve risk quantification efforts?

RISK RESPONSE PLANNING

R110 Are risks considered as they arise?

R111 Do teams seldom determine mitigation strategies or plan for contingencies for future risk events?

R223 Do project teams informally think about their strategy for dealing with future risk events and discuss strategies among themselves?

R224 Do the strategies include avoiding the risk, mitigating it or accepting it altogether?

R225 Does the organization have project teams develop a risk management plan that documents the procedures that will be used to manage risks?

R226 Does the plan cover things such as who is responsible, how the information will be maintained, how plans will be implemented and how reserves will be distributed?

R227 Does the plan provide for integration and consolidation into the project plan?

R228 Do large, highly visible projects develop contingency plans for near-term risks and mitigation strategies for all risk areas of concern?

R314 Does the risk response development process include templates for the risk management plan?

R315 Does your project team have contingency plans and mitigation strategies identified for each risk item?

R316 Does the organization allocate project reserves to cover contingency plans and mitigation strategies?

R417 Are all processes in place, documented and being used?

R418 Is the risk response development process fully integrated with cost management, time management, finance/accounting and strategic planning processes, and the project office?

R419 Are best practices and lessons learned being captured and made available to other projects?

R515 Is there an improvement process in place to continuously improve the risk response development process?

R516 Is there an improvement process in place to develop a risk management plan?

R517 Are best practices and lessons learned being used to improve Risk Response Planning?

R518 Is use of project reserves included in the determination of project efficiency and effectiveness?

R519 Is there a process in place for tracking the use of project reserves?

R520 Does the process for tracking the use of project reserves support management decisions during project execution?

R521 Are lessons learned being used to improve the development effort of identifying risk strategies?

RISK MONITORING AND CONTROL

R112 Is problem-solving done on a day-to-day basis versus a systematic approach to risk management and risk response strategies?

R229 During project execution, do individual project teams and segments of the organization apply their own approach to managing and controlling risks?

R230 Do project teams typically assign responsibility for each risk item as it occurs?

R231 Are risks discussed in team staff meetings and documented?

R232 Is management informed about the risk status of large, highly visible projects?

R233 Is there a process in place (developed and documented) to report risk status to key stakeholders?

R234 Does the process include a risk log that identifies risk items, who is responsible, the potential impact, the probability of impact, the mitigation strategy and the current status?

R235 Does your project team hold periodic meetings specifically for risk discussions to track and manage the risks and to discuss current status, workarounds and corrective actions?

R236 Does the organization collect metrics to track things such as the number of risk items and levels of concern?

R237 Is the risk status distributed to key stakeholders?

R238 Is the risk status incorporated into the project schedule?

R317 Is there a fully developed and utilized process for managing and controlling risks?

R318 Are project risks actively and routinely tracked?

R319 Are corrective actions taken, the risk management plan updates as risk events take place and/or things change, and project plans adjusted?

R320 Are metrics collected, analyzed and expanded to include success rate at mitigating risks?

R421 Are all processes in place, documented and being used?

R422 Is the risk control system fully integrated with the organization's control systems, monitoring programs, cost management and time management processes?

R423 Are best practices and lessons learned being captured and made available to other projects?

R522 Is there an improvement process in place to continuously improve the risk control process?

R523 Are best practices and lessons learned being used to improve Risk Monitoring and Control?

R524 Are risk assessments incorporated and included in the determination of project efficiency and effectiveness?

R525 Is there a process in place (developed and documented) utilizing risk assessments and the current risk status for management decisions during project execution?

R526 Are lessons learned being used to improve the monitoring and control efforts?

PROJECT PROCUREMENT MANAGEMENT PROCESSES

PROCUREMENT PLANNING

P101 Is the project manager's approach to determining project requirements an informal activity?

P102 Does the project manager use an informal approach to purchasing goods and services?

P103 Does the project manager decide whether to make or buy the goods/services in question after receiving a project request?

P201 Is this done in the process of creating a statement of work/product description?

P202 Does the project manager (and possibly team members) go through an informal analysis whether to make or buy the goods in question with input from the client department?

P203 Is this analysis heavily dependent upon the scope statement?

P204 Is a recommendation made to organizational management?

P205 If organizational management makes the decision to proceed with buying goods/services, do the procurement organization and the project team jointly create a procurement management plan?

P206 Does this plan identify items such as how to procure, what, how much, when, what deliverables, the quality necessary for these deliverables (based upon the project's quality standards), and timing for these deliverables?

P301 Do the project team and procurement organization present a formal analysis/recommendation report to both organizational management and client management?

P302 Does the make/buy recommendation and decision take into account effects and ramifications in such areas as capacity of the organization, most economical method, economic factors, organization situations, etc.?

P303 Is the make/buy decision made jointly by organizational management and client management?

P401 Is the make/buy decision made by a team composed of the project manager, organizational management, client management and the purchasing department?

P402 Does the decision include organizational factors such as available production capacity in other parts of the organization, ramifications on other active projects, ramifications on other parts of the organizational environment, etc.?

P403 Is input requested from all areas affected by the project or its product's?

P404 Are best practices and lessons learned being captured and made available to other projects?

P501 Is procurement planning evaluated on a periodic basis?

P502 Are enhancements continuously incorporated into the procurement planning process?

P503 Is the make/buy decision evaluated based upon efficiency and effectiveness metrics?

P504 Are historical data about projected make/buy costs evaluated against actual costs to determine if the decisions were sound and should be repeated in the future?

P505 Is just-in-time procurement incorporated to expedite the procurement planning process and reduce inventory carrying costs for the organization?

P506 Are best practices and lessons learned being used to improve Procurement Planning?

SOLICITATION PLANNING

P104 Are project requisitions prepared in a way similar to the way the organization prepared documentation to acquire typical goods and services?

P207 Is the scope statement reviewed and changed if necessary?

P208 If the decision is to build the product internally, is project scope revisited by the scope management process?

P209 If the decision is to buy the goods/services in question from an outside vendor, does the procurement organization take the lead on deciding which vendor(s) to contact?

P210 Does the procurement organization take the lead on preparing the procurement documentation with limited participation from the project team?

P211 Does the project team become more involved in the preparation of the procurement documentation?

P212 Is there a process in place for identifying contract requirements, identifying potential vendors, selecting the appropriate contract type, determining the best procurement approach, and developing procurement documentation?

P213 Does the organization establish clear evaluation criteria for use during proposal evaluation?

P214 Has the organization defined different procurement approaches (unilateral purchase orders; bilateral requests for quotation, request for proposal and invitation to bid) that are used by the project teams?

P215 Are purchase orders (POs) used when routine, standard cost items are needed?

P216 Are POs sent to the vendor with the expectation that it will accept it automatically?

P217 Are requests for quotation (RFQs) used when the goods and services are of relatively low dollar value, such as supplies and materials?

P218 Are requests for proposal (RFPs) used when there is a high dollar value involved and the goods and services are not standardized (and the project team has developed the product requirements but not the detailed design)?

P219 Is an invitation to bid used for high dollar value, "standard" goods and services (and the project team has a thoroughly documented design of the product which is conveyed to the potential vendor)?

P220 Do large, highly visible projects use the RFP process to obtain contract services for projects?

P304 Has the organization developed an expeditions process to access vendors/contractors (like a preferred bidders list)?

P305 Do vendor recommendations from the project team come from the preferred bidders list?

P306 Are vendor recommendations from the project team added to the preferred bidders list?

P307 Does the process for developing procurement documentation include procurement templates such as a statement of work format, status reporting, and other common procurement attachments?

P308 Has the organization identified and defined the types of contracts (firm-fixed-price, fixed-price-plus-incentive-fee, cost-plus-incentive-fee, cost-plus-fixed-fee, cost-plus-percentage-of-cost)?

P309 Is firm-fixed-price used when reasonably definite product specifications are available and costs are relatively certain (the team has done a detailed job of designing the product)?

P310 Is fixed-price-plus-incentive-fee used when the contract is for a substantial sum and involved a long delivery period?

P311 Is cost-plus-incentive-fee used when the contract involves a long delivery period with substantial amount of hardware development and test requirements?

P312 Is cost-plus-fixed-fee used primarily for research projects where the effort required for successful completion is uncertain until well after the signing of the contract?

P313 Is cost-plus-percentage-of-cost rarely used?

P405 Are the project's requisitions fully integrated with the organization's requisition process?

P406 Are best practices and lessons learned being captured and made available to other projects?

P507 Is the requisition process evaluated on a periodic basis?

P507 Are enhancements continuously incorporated into the requisition process?

P508 Is the requisition process automated and triggered by a project manager request?

P509 Is the requisition process evaluated based upon efficiency and effectiveness metrics?

P510 Does the organization have preferred contract vehicles?

P511 Does the organization have a list of preferred vendors who can respond to the requisition process expeditiously?

P512 Are best practices and lessons learned being used to improve Solicitation Planning?

SOLICITATION

P105 Does the organization follow an informal process of contacting several vendors, shopping and comparing prices?

P221 Does the procurement organization contact the vendor(s) and conduct price comparisons?

P222 Is the vendor(s) asked to commit to the final delivery date for the services with key milestones?

P223 Is the vendor(s) asked for a project plan at an appropriate level of detail?

P224 Is the quality to be met in the product specified to the vendor, based upon the project's quality standards?

P225 Are both the project manager and the purchasing department involved in the evaluation of the bids/proposals, using the established evaluation criteria?

P226 Is a process defined to solicit information from the industry?

P227 Is this process used on large, highly visible projects?

P314 Are contractors/vendors asked to comply with applicable project management processes and structure that is standard throughout the organization?

P315 Are vendors asked to supply a detailed plan, including a WBS and a detailed, sequenced activity list, in line with the project's structure?

P316 Is solicitation carried out jointly by the purchasing department and the project team, with input from the legal department?

P407 Is solicitation for the project fully integrated with the organization's solicitation process?

P408 Are best practices and lessons learned being captured and made available to other projects?

P513 Is solicitation evaluated on a periodic basis?

P514 Are solicitation enhancements continuously incorporated into the solicitation process?

P515 Are contractors evaluated based on efficiency and effectiveness metrics with regard to project performance?

P516 Do the project manager and the project team evaluate the vendors at the end of the project in terms of effectiveness, efficiency, responsiveness, timeliness and quality of product/service?

P517 Are the results of vendor evaluation fed back into the process and measured against the preferred providers list?

P518 Are best practices and lessons learned being used to improve Solicitation?

SOURCE SELECTION

P106 Evaluating industry information, negotiating the contract and finalizing the contract is not standardized?

P228 Is this process used on large, highly visible projects?

P317 Is source selection for the project fully integrated with the organization's solicitation process?

P409 Is it possible for the organization to leverage numerous requests to a given vendor and take advantage of economies of scale?

P410 Are best practices and lessons learned being captured and made available to other projects?

P519 Is source selection evaluated on a periodic basis?

P520 Are source selection enhancements continuously incorporated into the solicitation process?

P521 Are best practices and lessons learned being used to improve Source Selection?

CONTRACT ADMINISTRATION

P107 Are contracts loosely managed with minimal reporting requirements delineated within the contract?

P108 Are vendors/contractors managed to end dates only?

P229 Does the vendor supply periodic status reports to the project manager that reflect progress toward meeting key milestones?

P230 If changes to the plan represent a scope change, is the normal change management process within the project integration process used for large, highly visible projects?

P231 Does the vendor report on a regular basis on the reporting level agreed upon in the contract?

P232 Is the format of the vendor's report specified in the procurement process?

P233 Is the reporting frequency per the procurement management plan and contract?

P234 Is information on work results provided to the project integration management process for internal progress reporting?

P235 If there is a change to the plan, does the project manager send corrective action information to the project integration management process, resulting in plan updates or use of the change control process?

P318 Does the vendor report in a timely manner to the project manager against progress on the detailed sequenced activity list?

P319 Are any changes/issues immediately communicated to the project manager?

P320 Does the project manager forward changes/issues to the change management process, which is fully implemented within the project?

P321 Is the client integrally involved in testing of the product?

P322 Does the client sign off on contract completion?

P323 Does the project manager sign off on contract completion?

P411 Is the vendor required to report progress against plan using the organization's standard project management tools and techniques?

P412 Are weekly status reports provided from the vendor to the project manager?

P413 Are the weekly status reports integrated into the organization's standard status reporting mechanisms?

P414 Are vendors more integrated into the project planning process?

P415 Are best practices and lessons learned being captured and made available to other projects?

P522 Is contract management evaluated on a periodic basis?

P523 Are enhancements to contract management continuously incorporated?

P524 Does the organization consider strategic alliances with preferred vendors?

P525 Do both organizations adhere to a high standard of project performance and quality in their products/services?

P526 Are lessons learned (about the procurement process within the project with regard to effectiveness and efficiency) captured on large, highly visible projects?

P527 Does a performance database exist to capture performance information on the vendors/contractors?

P528 Are best practices and lessons learned being used to improve Contract Administration?

CONTRACT CLOSE-OUT

P109 Is acceptance and contract closure an informal process?

P236 Is a process established and documented for formal acceptance and contract closure?

P324 Is closure information and formal acceptance provided to the communication process?

P416 Is contract close-out evaluated on a periodic basis?

P417 Are best practices and lessons learned being captured and made available to other projects?

P529 Are enhancements to contract close-out continuously incorporated?

P530 Are best practices and lessons learned being used to improve Contract Close-Out?

About the Author

Robert K. Wysocki, Ph.D., has more than 38 years of experience as a project management consultant and trainer, information systems manager, systems and management consultant, author, training developer, and provider. He has written 13 books on project management and information systems management. One of his books, *Effective Project Management, Third Edition*, has been a best seller and is recommended by the Project Management Institute for the library of every project manager. He has more than 30 publications and presentations in professional and trade journals and has made more than 100 presentations at professional and trade conferences and meetings. He has developed more than 20 project management courses and trained over 10,000 project managers.

In 1990 he founded Enterprise Information Insights, Inc. (EII), a project management consulting and training practice specializing in project management methodology design and integration, project support office establishment, the development of training curriculum, and the development of a portfolio of assessment tools focused on organizations, project teams, and individuals. His clients include AT&T, Aetna, Babbage Simmel, British Computer Society, Boston University Corporate Education Center, Computerworld, Converse Shoes, the government of the Czech Republic, Data General, Digital, Eli Lilly, Harvard Community Health Plan, IBM, J. Walter Thompson, Peoples Bank, Sapient, The Limited, the state of Ohio, Travelers Insurance, and several others.

He is the Effective Project Management Series Editor for Artech House. He is a member of the ProjectWorld Executive Advisory Board, the Project Management Institute, the American Society of Training & Development, and the Society of Human Resource Management. He is past association vice president of AITP (formerly DPMA). He earned a B.A. in mathematics from the

University of Dallas, and an M.S. and a Ph.D. in mathematical statistics from Southern Methodist University.

Index

Recent Titles in the Artech House Technology Management and Professional Development Library

Bruce Elbert, Series Editor

Advanced Systems Thinking, Engineering, and Management, Derek K. Hitchins

Critical Chain Project Management, Lawrence P. Leach

Decision Making for Technology Executives: Using Multiple Perspectives to Improve Performance, Harold A. Linstone

Designing the Networked Enterprise, Igor Hawryszkiewycz

Engineering and Technology Management Tools and Applications, B. S. Dhillon

The Entrepreneurial Engineer: Starting Your Own High-Tech Company, R. Wayne Fields

Evaluation of R&D Processes: Effectiveness Through Measurements, Lynn W. Ellis

From Engineer to Manager: Mastering the Transition, B. Michael Aucoin

Introduction to Information-Based High-Tech Services, Eric Viardot

Introduction to Innovation and Technology Transfer, Ian Cooke and Paul Mayes

ISO 9001:2000 Quality Management System Design, Jay Schlickman

Managing Complex Technical Projects: A Systems Engineering Approach, R. Ian Faulconbridge and Michael J. Ryan

Managing Engineers and Technical Employees: How to Attract, Motivate, and Retain Excellent People, Douglas M. Soat

Managing Successful High-Tech Product Introduction, Brian P. Senese

Managing Virtual Teams: Practical Techniques for High-Technology Project Managers, Martha Haywood

Mastering Technical Sales: The Sales Engineer's Handbook, John Care and Aron Bohlig

The New High-Tech Manager: Six Rules for Success in Changing Times, Kenneth Durham and Bruce Kennedy

Planning and Design for High-Tech Web-Based Training, David E. Stone and Constance L. Koskinen

Preparing and Delivering Effective Technical Presentations, Second Edition, David Adamy

Reengineering Yourself and Your Company: From Engineer to Manager to Leader, Howard Eisner

The Requirements Engineering Handbook, Ralph R. Young

Running the Successful Hi-Tech Project Office, Eduardo Miranda

Successful Marketing Strategy for High-Tech Firms, Second Edition, Eric Viardot

Successful Proposal Strategies for Small Businesses: Using Knowledge Management to Win Government, Private Sector, and International Contracts, Third Edition, Robert S. Frey

Systems Approach to Engineering Design, Peter H. Sydenham

Systems Engineering Principles and Practice, H. Robert Westerman

Systems Reliability and Failure Prevention, Herbert Hecht

Team Development for High-Tech Project Managers, James Williams

Recent Titles in the Artech House Technology Management and Professional Development Library

Bruce Elbert, Series Editor

Advanced Systems Thinking, Engineering, and Management, Derek K. Hitchins

Critical Chain Project Management, Lawrence P. Leach

Decision Making for Technology Executives: Using Multiple Perspectives to Improve Performance, Harold A. Linstone

Designing the Networked Enterprise, Igor Hawryszkiewycz

Engineering and Technology Management Tools and Applications, B. S. Dhillon

The Entrepreneurial Engineer: Starting Your Own High-Tech Company, R. Wayne Fields

Evaluation of R&D Processes: Effectiveness Through Measurements, Lynn W. Ellis

From Engineer to Manager: Mastering the Transition, B. Michael Aucoin

Introduction to Information-Based High-Tech Services, Eric Viardot

Introduction to Innovation and Technology Transfer, Ian Cooke and Paul Mayes

ISO 9001:2000 Quality Management System Design, Jay Schlickman

Managing Complex Technical Projects: A Systems Engineering Approach, R. Ian Faulconbridge and Michael J. Ryan

Managing Engineers and Technical Employees: How to Attract, Motivate, and Retain Excellent People, Douglas M. Soat

Managing Successful High-Tech Product Introduction, Brian P. Senese

Managing Virtual Teams: Practical Techniques for High-Technology Project Managers, Martha Haywood

Mastering Technical Sales: The Sales Engineer's Handbook, John Care and Aron Bohlig

The New High-Tech Manager: Six Rules for Success in Changing Times, Kenneth Durham and Bruce Kennedy

Planning and Design for High-Tech Web-Based Training, David E. Stone and Constance L. Koskinen

Preparing and Delivering Effective Technical Presentations, Second Edition, David Adamy

Reengineering Yourself and Your Company: From Engineer to Manager to Leader, Howard Eisner

The Requirements Engineering Handbook, Ralph R. Young

Running the Successful Hi-Tech Project Office, Eduardo Miranda

Successful Marketing Strategy for High-Tech Firms, Second Edition, Eric Viardot

Successful Proposal Strategies for Small Businesses: Using Knowledge Management to Win Government, Private Sector, and International Contracts, Third Edition, Robert S. Frey

Systems Approach to Engineering Design, Peter H. Sydenham

Systems Engineering Principles and Practice, H. Robert Westerman

Systems Reliability and Failure Prevention, Herbert Hecht

Team Development for High-Tech Project Managers, James Williams

For further information on these and other Artech House titles, including previously considered out-of-print books now available through our In-Print-Forever® (IPF®) program, contact:

Artech House
685 Canton Street
Norwood, MA 02062
Phone: 781-769-9750
Fax: 781-769-6334
e-mail: artech@artechhouse.com

Artech House
46 Gillingham Street
London SW1V 1AH UK
Phone: +44 (0)20 7596-8750
Fax: +44 (0)20 7630-0166
e-mail: artech-uk@artechhouse.com

Find us on the World Wide Web at:
www.artechhouse.com